BASIC

and

Chemistry

BASIC
and
Chemistry

New Impression

Leonard Soltzberg, Simmons College

Arvind A. Shah, Babson College

John C. Saber, Babson College

Edgar T. Canty, Babson College

Houghton Mifflin Company   Boston

Atlanta   Dallas   Geneva, Illinois   Hopewell, New Jersey   Palo Alto   London

Printed in the U.S.A.

Library of Congress Catalog Card Number: 74-15587

ISBN: 0-395-21720-2

Contents

Correlation Chart

CHAPTER	GENERAL CHEMISTRY								PHYSICAL CHEMISTRY		
TEXT	Becker & Wentworth	Mahan (College Chemistry)	Masterson & Slowinski	Mortimer	Nebergall et al. (College Chemistry)	Nebergall et al. (General Chemistry)	Quagliano & Vallarino	Sienko & Plane 4th ed.	Barrow 3rd ed.	Castellan 2nd ed.	Moore 4th ed.
6 Units	App. A, B, G	-	1	1	1, App. A	1, App. A	2	App. 2	-	-	-
7 Stoichiometry	3	1	2,3	5	2,9	2,9	4	5	-	1	-
8 Gases	2,11	2,9	5	6	8	8	8,9	6	1,2	2,3,4	1,4
9 Solutions	2,12	4	11	7,9	12	12	18,19	10	21	5,13	7
10 Equilibrium	13	5,6	12,16 17,18	14,15 16,17	17,20 30,31	17,20 30,31	20,24, 25	12	9	11	8
11 Radiochemistry	17	17	23	21	28	28	-	2	-	31	-
12 Molecular Structure Determination	-	3	10	7	-	-	10	7	13,14	26	15,17,18
13 Atomic Theory	4,6	10,11	7,8	2,3	4	4	15	3,4	3,11,12	19,22,23	13,14,15
14 Thermodynamics and Least Squares	10, App. F	8,9	13, App. 1	13	17	17	17,26	9,11	6,7,9 17,20	11,12,32	8,9

vi

Preface to Teachers

This book gives students of chemistry an opportunity to solve problems of chemical importance in a computer environment. It attempts to make three main points to the student:

1. Quantitative investigation in science can yield important insights.

2. The computer can aid a quantitative investigation in several specific ways.

3. Certain kinds of problems are appropriately solved using computers, while others are not.

We establish these ideas by leading the student through many of the quantitative problems encountered in introductory chemistry, illustrating the solution of these problems by computer and asking the student to apply computer methods to the solution of new problems. Since not all problems are best solved by computer, inappropriate problems are not provided. This volume is not a comprehensive problem book in the traditional sense, although most of the quantitative concepts associated with introductory chemistry are covered. It is intended that the student will not only acquire the ability to solve specific chemical problems but will also develop a feel for when to use the computer in his work.

The first part of the book provides the necessary BASIC programming apparatus. Chapters 1 through 5 take the student in stages from turning on the teletypewriter, through elementary problems, to detailed analysis and the concept of using a flow chart. Then this part helps the student to apply this knowledge to complex problem solving. We recommend that the student have access to a teletypewriter to use in conjunction with the book. As with learning any new skill, the more the student practices, the better he will understand the subject matter.

The second part deals with chemistry. The exercises in chapters 6 and 7 (on units and on stoichiometry) are designed to help the student develop programming competence. Later chapters involve more imaginative computer uses, sometimes calling upon the student to put together a computer investigation of a chemical system from program segments provided in the book. Because of the unique emphasis of this book, it is seasoned with "how to" tips, both on solving chemistry problems and on surmounting the associated programming hurdles.

The sample programs in the text stress getting-the-job-done, rather than elegant programming. The BASIC programming language was chosen for the book since it is the easiest language for a beginner to learn, and since it is widely accepted in academic circles. Once the student has mastered the concepts of programming in BASIC, he can with relative ease make the transition to other programming languages, such as FORTRAN.

While the book is basically intended for the general chemistry course, many of the exercises--particularly the simulations--would be instructive for more advanced students, for example in physical chemistry. Thus, the book can be used by the student and teacher for periodic reference throughout the undergraduate chemistry program.

The professional value of integrating computer awareness into the modern undergraduate chemistry curriculum is self-evident. In addition, experience has revealed two other benefits accruing from curricular computer use. The first is that in writing a computer program to solve a problem, the student must in essence "teach" the computer how to do the problem. The precision of thought required for this process generally insures that the student who can do the exercise has a good understanding of the problem; this is not necessarily true with conventional hand-solved problems. Secondly, the excitement associated with using a computer seems to enhance student enthusiasm for the subject at hand.

We hope that this book will help your students acquire both a better understanding of chemistry and a new tool for extending that understanding.

Introduction

The Computer in Science

Welcome to the realm of science. The chemistry course on which you have embarked can be a journey during which you will acquire heightened awareness of the beauty of nature. The scientist's fundamental motivation is a fascination with the unique creations and delicate interrelationships of the natural world, and we hope to share this fascination with you.

In observing nature's ways, one is often first struck by a qualitative image: a pebble falls to the earth when released; an iron bar glows when heated; a cup of tea lightens in color when lemon is added. Much of our present understanding of nature began with qualitative observations. However, the curious mind stimulated by such events begins to formulate more probing questions. How fast does the pebble fall? What temperature is the iron bar when it starts to glow? How much lemon is needed to lighten the tea? These questions demand quantitative answers, which must be sought through measurement and calculation.

Thus it is that any branch of science, as it matures, displays increasing concern with quantitative questions. The scientist seeks ways of making increasingly precise measurements: new instruments are created. Then he seeks better ways of handling these measurements, of doing calculations--and here, the scientist turns to the computer.

The computer has many roles in modern science. Chemists, anthropologists, physicists, biologists, psychologists--people in virtually every branch of contemporary science--are making important use of computers. These applications include such varied adventures as controlling a manned spacecraft or elucidating the nature of visual perception. In spite of the mind-stretching variety of tasks to which science sets the computer, essentially all of these applications involve one, or a combination, of the four following jobs.

1. Calculation or "number crunching." Here, the computer is basically used as a calculating machine, immensely faster and more reliable than its human operator.

2. Information storage and retrieval. The computer is used as a library of information that can be accessed with great speed and indexed according to many different sets of criteria. This type of computer use enables a person to discern patterns or relationships among data that would not be apparent from a simple visual scanning of the data.

3. Simulation. The computer is made to "behave like" some system that cannot be studied directly because of inaccessibility, danger, time limitations, etc.

4. Instrument control. Here, a computer relieves the scientist of the time-consuming routine operation of data-gathering instruments by automatically controlling the instruments and storing the data produced.

The purpose of this book is to introduce you to the exciting world of computers and to help you use the computer in your study of chemistry. The first part of the book shows you the essentials of programming, and the rest of the book will guide you in using your new skill to solve chemistry problems.

Before we begin, there are two cautions you should heed. The first is that not every scientific problem needs to be solved on a computer. There are some jobs that are too trivial to employ a computer; and there are jobs that are done better by human beings, such as those demanding judgment or inspiration. The kinds of problems in this book for which we shall use the computer are those involving:

1. doing a calculation over and over with different data

2. doing a calculation requiring a large amount of data

3. evaluating a complex formula

4. carrying out a multi-step procedure

5. storing and retrieving information

6. simulating a system

In all of these cases, the computer's speed, reliability, storage capacity, and patience make its use advantageous.

The second caution relates to the perennial question of significant figures. This can be a pitfall even with hand calculation; but since the computer always displays answers with six or more digits, it is essential to bear in mind that the number of meaningful digits in an answer is limited by the least precise piece of data used in the calculation.

Now, let's begin!

1

Getting Started

In this chapter we will discuss the general concept of a computer and the BASIC language in particular. Next, we will explain the concept of computer time-sharing, and finally, we will distinguish between "system commands" and "BASIC commands".

1.1 The Magnificent Bully

The computer, regardless of the advertising and preconceived ideas that we may have, is an obedient slave, albeit a very literal one. One of the hardest realizations that we will have to reach is that the computer is only as smart as we are, but is infinitely faster. Therein lies one of the reasons for characterizing the computer as the "magnificent bully". Its speed has enabled business and science to keep pace with the expanded needs of our society to compute and keep track of the current state of affairs. For example, in the last ten years the population has increased from 180 million to 200 million. In the same period, the total of man's experience recorded on paper has doubled. Each of the lunar missions required more computation than all of the world's population could have accomplished in its life-time. Without the aid of the magnificent bully, as we call the computer, we would have been hard pressed to cope with the simultaneous explosions of population and knowledge that have occurred.

The computer is magnificent because it not only stores billions of pieces of

data that are readily available for use, but it also acts as a teacher, lawyer,
doctor, etc. The potential for new uses is still expanding as men and women are
able to recognize how old problems can be solved by using the tool we call the
computer. The computer is a bully because, more than any other device we use, it
requires the exact specification of what we want it to do. There are no compro-
mises because it will do only what we direct it to do, regardless of whether it is
what we want it to do. There is the rub: the difference between directing and
wanting.

One reason that this magnificent bully acts as it does is that it is nothing
more than a high-speed electronic device which processes symbols and produces sym-
bolic results. To appreciate this, we must realize that a computing system is
usually divided into two parts: the physical apparatus called the hardware, and the
programming instructions called the software. Although we need not understand the
hardware completely, we should be aware that there are four parts common to any
computer. The first is the arithmetic and control unit which performs the actual
calculations and supervises the operation of the other parts. The second is the
memory unit(s) used to store information. The third and fourth are the input and
output units which are used to get information into and out of the computer. It is
important to realize that the computer has shrunk from a machine 900 feet long, 3
feet wide, and 8 feet high when it was introduced in the 1940's to specialized computers today which can be placed on a desk.

The software is made up of two major parts. The first is the operating system
which uses the control unit to direct the performance of the rest of the hardware.
In the beginning, this was the only means by which the user could direct the computer, and hence most systems were so complex that it took a skilled engineer to
even hope to get the machine to function properly. Although this aura of diffi-
culty hangs on, it is no longer merited because of the advent of the second part of
the software called the "compiler". The compiler is nothing more than an automatic
translator which allows the user to write instructions to the computer in an
English-like language, and then translates these instructions to "machine
language"--a form which the machine can utilize.

Now there are almost as many languages and dialects for computers as there are
for humans. The language which we discuss in this book is called BASIC, which is

an acronym standing for <u>B</u>eginner's <u>A</u>ll-purpose <u>S</u>ymbolic <u>I</u>nstruction <u>C</u>ode.   It was

developed at Dartmouth College in the early 1960's by Dr. John Kemeny and

Dr. Thomas Kurtz.   Since that time it has grown in popularity to become one of the

major computer languages because it is so easy to learn and apply to problems.

Moreover, once the user has mastered BASIC, he will find that the concepts he has

learned enable him to understand more sophisticated languages such as FORTRAN,

COBOL, and APL.

     There are many dialects of BASIC in current use.   It is therefore desirable

for the user to check with his computer center to see which dialect it uses.   The

BASIC described in this book corresponds to the most common usage of the language,

and the programs shown in this book were run on a Hewlett-Packard 2000A.   A table

comparing the various BASIC commands for several other computers is given in

Appendix 2.

1.2   Computer Time-sharing

     Before going further a word should be said about time-sharing.   It was in the

early 1960's that the idea of sharing the cost and benefits of computer facilities

arose out of the need to keep a computer busy with work.   At the same time, com-

puter scientists began to realize that the computer was so fast that it could be

used almost simultaneously by many users.   Thus the idea of computer time-sharing

proved to be a good one, and today thousands of businessmen, scientists, engineers,

and students, all who normally would not be able to afford a computer of their own,

have the advantage of using an economical, fast computer.   Although it appears that

each individual using a terminal is the sole user, in actuality a computer in a

time-sharing system can handle scores of other users all at the same time.

1.3   System Commands and BASIC Commands

     In general, the commands to the computer can be divided into two types:

<u>system</u> <u>commands</u> and <u>BASIC</u> <u>commands</u>.   System commands are intended to be executed

immediately and effect machine operations such as starting, stopping, and execution

of the program.   (A program, by the way, is a sequence of instructions telling the

computer to perform a specific task.)   System commands are discussed in Chapter 2.

BASIC commands relate to the way we want the machine to perform the actual calcula-

tions whenever we run the program.   These are discussed in Chapters 3, 4, and 5.

2
Talking to the Computer via the Teletypewriter — System Commands

In this chapter, we will discuss the system commands which are used to control the operation of the computer. First, we explain how to initiate communication with the computer (sign on), and how to end this communication (sign off). Next, we show how to run prewritten programs, how to display any such program, and also how to interrupt the computer during an operation. Finally, we discuss the use of the paper tape device and some other system commands.

2.1 Signing On and Off--HEL and BYE

There are many devices that we can use to communicate with the computer; indeed the ability of the computer to understand commands has been extended to accepting spoken commands. We will confine ourselves in this book, however, to the most commonly used device in time-sharing, the teletypewriter, shown in Figure 2.1. Note that the teletypewriter is really made up of three separate components, namely the keyboard in the center, the paper tape device on the left, and the audio coupler on the right. The audio coupler is used to set up the communication link with the computer and is turned on first. Next we turn the knob on the right-hand side of the teletypewriter below the keyboard to the LINE position and dial the telephone number of the computer center. When we hear a high-pitched squeal we put the telephone handset into the coupler, thus completing the connection to

4

Figure 2.1

the computer. Now as we type on the keyboard, the teletypewriter emits an electronic signal, which is converted by the audio coupler into a form which can be
sent over the telephone lines.

Next we are ready to sign on the computer, i.e., identify ourselves as
authorized users. We do this by typing on the keyboard shown in Figure 2.2 the
system command HELLO, or the alternate form HEL (since the computer looks at only
the first three letters of any system command), and a password such as B∅∅1,FLIP.
(NOTE: We are adopting the convention of using "∅" to represent a zero and thus
distinguish it from the letter oh "O".) The typed line will appear on the teletypewriter as

HEL-B∅∅1,FLIP

after which we now depress the RETURN key. The RETURN key must be depressed after
each line so that the computer knows that the line has been completed. If all has
gone well, the computer will now respond by returning the printing head, advancing
the line, and typing a message such as

"READY"

or

"WELCOME TO THE SYSTEM"

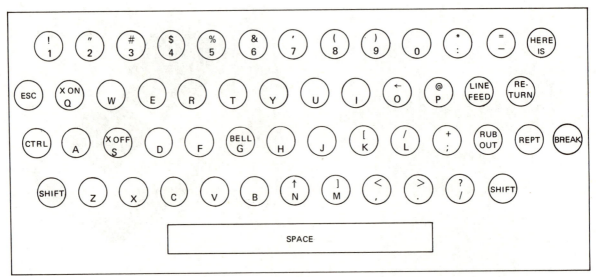

Keyboard

Figure 2.2

to show us that we have been accepted as a legitimate user.

In the event that we made an error in typing, the following two general techniques may be used to make corrections.

Technique 1

Depress the back arrow (←) key, usually the upper case letter "O" (oh) as can
be seen in Figure 2.2, for each character to be deleted. Then retype the line
correctly.

Example 2.1

HEL-BX←∅∅1,FLIP is read as HEL-B∅∅1,FLIP

or

HEL-BZPL←←←115,NED is read as HEL-B115,NED

Technique 2

Depress the ESC (ESCAPE) key in the upper left keyboard area, also to be seen
in Figure 2.2. This will delete the entire line, automatically return the printing
head, and advance the line. Then we can retype the line correctly. (Different
terminals have other keys to accomplish the same purpose.)

Once we have completed our work we sign off by typing the system command

BYE

(NOTE: Remember to depress the RETURN key at the end of every line.) At this
point the computer will print the amount of time used on the terminal.

Example 2.2

BYE

∅87 MINUTES OF TERMINAL TIME

Now we hang up the telephone and turn off the audio-coupler and the teletypewriter.

2.2 Running a Prewritten Program--GET and RUN

After signing on as indicated in the previous section, we are now ready to use
prewritten programs with no further knowledge of programming itself. These prewritten programs are divided into two categories: public library programs and user
library programs. Regardless of the password used in signing on, public library
programs are available to all users of the computer. To recall such a program from
the library, we simply type the system command GET-$ followed by the program name.

Example 2.3

GET-$DIAMND

where DIAMND is the name of the program we wish to execute. To actually execute
the program, we type the system command

RUN

In summary, we have the following displayed on the teletypewriter:

    User: HEL-B∅∅1,FLIP
    Computer: WELCOME TO THE ACCOMP SYSTEM
    User: GET-$DIAMND
    User: RUN

The computer will execute the prewritten instructions in DIAMND, and when program
execution is completed, the computer will print on the teletypewriter the message

DONE

as shown in Figure 2.3.

HEL-BØØ1,FLIP

WELCOME TO THE ACCOMP SYSTEM

GET-$DIAMND

RUN

DIAMND

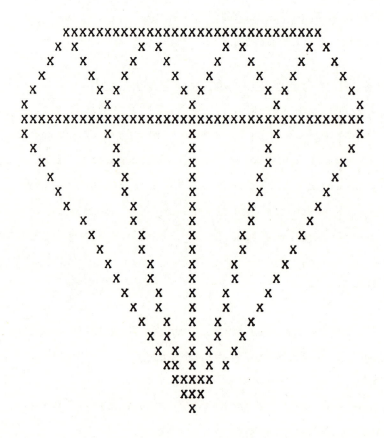

DONE
Figure 2.3

The user library programs are available only to those users who know the particular password(s) under which the programs are stored. Each password has access to its own private library as well as the public library. To recall a program stored in a user library, we type the system command GET- followed by the program name (note that there is no "$").

Example 2.4

        GET-MYOWN

where MYOWN is the name of the program we wish to execute. Now, as with a public library program, to actually execute MYOWN we simply type the system command

                                        RUN

        Finally, as noted in the previous section, once we have completed our work we sign off the computer by typing

                                        BYE

2.3 Displaying a Program--LIS

        If we have recalled a program as shown in the previous section or we have just written a program on the teletypewriter, we can display it simply by typing the system command LIS (LIST). This will cause the computer to print out on the teletypewriter a copy of the program as stored.

Example 2.5

        User: HEL-BØØ1,FLIP

        Computer: WELCOME TO THE ACCOMP SYSTEM

        User: GET-MYOWN

        User: LIS

        Computer: MYOWN

                    2Ø PRINT 8

                    3Ø PRINT 75

                    4Ø END

Although we will discuss this program in detail in the next chapter, we wish to point out that each line of the program above begins with a "line number". These line numbers can be used to display only part of a program by typing LIS- followed by the line number of the first line we wish displayed. The computer will then display this line and all following lines in the program.

Example 2.6

    User: GET-MYOWN
    User: LIS-3∅
    Computer: MYOWN
               3∅ PRINT 75
               4∅ END

Finally, suppose that we wish to stop the computer while it is listing or running a program. To do this we momentarily depress the BREAK key on the right-hand
side of the keyboard, as shown in Figure 2.2*. The computer will stop what it is
doing, automatically return the printing head, advance the line, and lastly print
the message STOP.

2.4 Using Paper Tape--TAP, KEY, and PUN

There are many ways in which the user can store programs and data external to
the computer. For example, punched paper tape, magnetic tape, and punched cards
may be used for such storage. The most commonly used method in time-sharing is
punched paper tape because of its simplicity and its compatibility with the typical
teletypewriter. For these reasons we will now discuss punched paper tape in
greater detail.

The primary reasons for using external storage are twofold. First, internal
storage is usually limited and always expensive. Second, time-sharing systems
usually charge for the time that the user is connected or signed on to the computer. Using paper tape, the user can enter his program or data at the rate of 10
characters per second which is much faster than even a trained secretary can type.

The paper tape device shown in Figure 2.1 is located to the left of the keyboard on the terminal. To operate this device while not being connected to the
computer (off-line rather than on-line as in Section 2.1) we turn the knob on the
right-hand side of the teletypewriter to the LOCAL position and depress the ON button on the paper tape device. Now, for ease in the handling of the paper tape, we
get a "leader" on the tape by simultaneously depressing the REPT (repeat) and
RUBOUT keys. Then we type our program or data as usual with the additional step of
ending each line by depressing the RETURN, LINE FEED, and RUBOUT keys in that

*NOTE: Do not depress the key for too long or else the telephone connection may be
        broken.

precise order.   These actions return the carriage, feed one line, and reset the
printing head to its normal position.   In the event that we made an error in
typing, we can correct it, as described in Section 2.1, by using the back-arrow or
the ESC key.   Lastly, on completing the program or data we get a "trailer" on the
tape by once more simultaneously depressing the REPT and RUBOUT keys.   Then we
turn off the teletypewriter and carefully tear off the tape.   Note that the tape
will have a ⟩ shape at the beginning and a ⟩ shape at the
end.

Example 2.7

   To punch the program MYOWN of the previous section on paper tape, we proceed
as follows:

   1. Turn the teletypewriter to LOCAL.

   2. Depress the paper tape device ON button.

   3. Simultaneously depress REPT and RUBOUT keys.

   4. Type: 2Ø PRINT 8

      Depress: RETURN, LINE FEED, and RUBOUT keys.

      Type: 3Ø PRINT 75

      Depress: RETURN, LINE FEED, and RUBOUT keys.

      Type: 4Ø END

      Depress: RETURN, LINE FEED, and RUBOUT keys.

   5. Simultaneously depress REPT and RUBOUT keys.

   6. Turn off the teletypewriter.

In order to read a tape into the computer we first sign on as in Section 2.1.
Next we place the tape into the tape reader with the pointed end toward the user
and the sprocket holes properly aligned.   Finally, we type the system command TAP
(TAPE) which transfers control from the keyboard to the paper tape device, and push
the lever on the paper tape device to the START position.   The computer will simultaneously read the tape and print the corresponding material on the terminal.

   After the tape has been read, we return control back to the keyboard by typing
the system command KEY.   If the program or data has been properly written in BASIC,
the user may proceed.   On the other hand, if there has been a syntax error (i.e.,
an error in the structure of a BASIC line), the computer will automatically respond

by typing a message (a diagnostic) indicating the nature of the mistake and that
the line has not been accepted. These errors can be corrected by properly retyping
the associated lines. Any other system command such as LIS or RUN will also return
control to the keyboard, but these will cause the computer to take other actions as
well.

Example 2.8

    Suppose MYOWN had been incorrectly punched as

    2∅ PRUNT 8

    3∅ PRINT 75

    4∅ END

    User: HEL-B∅∅1,FLIP

    Computer: WELCOME TO THE ACCOMP SYSTEM

    User: TAP

Place the tape in the reader and push up the START lever.

    Computer: 2∅ PRUNT 8

              3∅ PRINT 75

              4∅ END

    User: KEY

    Computer: MISSING ASSIGNMENT OPERATOR IN LINE 2∅

              LAST INPUT IGNORED, RETYPE IT

    User: 2∅ PRINT 8

    User: LIS

    Computer: 2∅ PRINT 8

              3∅ PRINT 75

              4∅ END

Note that we listed the program at the end to make sure that the correction was
made properly. This should always be done by beginning users.

    Finally, there are many times when we wish to record on tape either a program
or data stored in the computer. To do this we simply type the system command PUN
(PUNCH), turn on the paper tape punch, and depress the RETURN key. The computer
will automatically punch on paper tape and type on the terminal the requested
material stored in the computer.

2.5 Other System Commands

So far we have discussed the system commands HEL, BYE, GET, RUN, LIS, TAP, KEY, and PUN. A detailed list and description of other system commands for the Hewlett-Packard 2000 is given in Appendix 1.

3
Introduction to BASIC Commands

In this chapter we will discuss some of the fundamental BASIC commands. We will begin by introducing a simple program to illustrate in general how BASIC commands work. Next, we will present the arithmetic operations and the technique for storing information in the computer. Then we will discuss three alternate ways of getting information into the computer. Finally, we will demonstrate how to analyze a problem to be solved on the computer, and how to lay out a "road map" for the solution using the concept of flowcharting.

3.1 A Simple Program--REM, PRINT, and END

As we mentioned in Section 2.2, a program is a sequence of instructions telling the computer to perform a specific task. Each instruction generally specifies one operation to be performed, as shown in the following example.

Example 3.1

Line Number	Instruction
10	REM THIS IS A SIMPLE BASIC PROGRAM
20	PRINT 8
30	PRINT 75
40	END

14

This is a complete program consisting of four instructions. Note that each instruction is preceded by a line number which tells the computer where the instruction belongs in the sequence of operations. The line numbers themselves are in increments of ten to allow for later insertions or corrections if necessary, but in practice may be anything from 1 to 9999.*

Line 1∅ contains the REM (REMARK) instruction. This instruction is never executed by the computer but is used to give information about the program to a user. For example, in the above program it is used to identify the program. It is also helpful in defining what "variable names" represent, what various sections of a program are designed to accomplish, etc.

Line 2∅ contains the PRINT instruction. This instruction will have the computer print on the teletypewriter the number following the word PRINT. For example, in the above program, the computer will type the number 8 on the teletypewriter. This instruction can be used in many other ways, which will be described in later sections of this chapter.

Line 3∅, as line 2∅, contains a PRINT instruction. In this case, the computer will print the number 75 on the teletypewriter.

Lastly, line 4∅ contains the END instruction. Every BASIC program must have one END instruction, and this instruction must have the highest line number in the program. The reason for this is that the END instruction signifies to the computer that it has the complete program. Note that as with the REM statement, the END instruction does not cause any specific action to be taken in the program.

Now, if the user wants to have the computer actually execute or run the above program, he proceeds as follows:

1. Sign on the computer (Section 2.1).
2. Type the program on the teletypewriter as written. (We will show how to correct errors in the next section).
3. Type the system command RUN.
4. Sign off when the execution of the program is completed. (Section 2.1.)

Thus, we have

*NOTE: Some computers may allow numbers from 1 to 99999.

```
User:        HEL-BØØ1,FLIP

Computer:    WELCOME TO THE ACCOMP SYSTEM

User:        1Ø REM THIS IS A SIMPLE BASIC PROGRAM

             2Ø PRINT 8

             3Ø PRINT 75

             4Ø END

             RUN

Computer:    8

              75

             DONE

User:        BYE

Computer:    ØØ1 MINUTES OF TERMINAL TIME
```

Finally we wish to point out that the number "8" and the "7" in the number "75"
were both printed in the second column. The reason for this is that the first column is reserved for the algebraic sign associated with numeric data. When the number is positive, the sign is omitted and a blank character is printed; when the
number is negative, the sign is printed.

3.2 Correcting Errors and Changing a Line

If the user detects an error while typing a line in his program before he de-
presses the return key, then he has the same two techniques for correcting the
error that we discussed in Section 2.1. That is, he may either use the back arrow
(←) to delete one or more characters, or he may depress the ESC (Escape) key or its
equivalent to delete the entire line.

On the other hand, if the user has made an error, has not detected it, and has
depressed the RETURN key, then one of two things will happen. First, if the error
is a violation of the structure of BASIC the computer will automatically respond by
typing a diagnostic message.*

Example 3.2

Suppose the user typed
2Ø PRUNT 8

*NOTE: The user should check the system he is using for specific error correction
        techniques.

The computer will respond by typing

    ERROR

If the user does not recognize the error which he made, he may type any letter,
such as X. The computer will respond as shown below by identifying the error.

    ERROR X MISSING ASSIGNMENT OPERATOR

At this point, the user simply retypes the line correctly as

    2∅ PRINT 8

Second, if the error is a logical one in the structure of the program, the
computer will not ordinarily be able to recognize it. During the program run, the
user usually detects this error, and corrects it by retyping the incorrect line(s).

Example 3.3

    Suppose we typed

    1∅ REM THIS PROGRAM DEMONSTRATES ERROR CORRECTION

    2∅ PRINT 88

    3∅ END

    RUN

The computer would respond by typing

     88

    DONE

If we had wanted the computer to print 23 instead of 88, we could erase the old
line 2∅ by simply retyping it as

    2∅ PRINT 23

Now if we run the program, we would get

    RUN

     23

    DONE

This example demonstrates the general technique of changing a line in a program.
Finally, sometimes we want to delete the entire line rather than change it to something else. In this case we simply type the line number alone, i.e., with no BASIC
instruction.

3.3 More on PRINT--Quotes, Commas, Semicolons, and TAB

    The PRINT statement can be used not only to print out numbers but also any

statement consisting of letters, numbers, and symbols, called alphanumeric or
alphameric information. All we have to do is enclose the required statement in
quotation marks.

Example 3.4

        10 PRINT "THIS IS MY FIRST PROGRAM"

        20 PRINT "MY NAME IS RUMPLESTILTSKIN AND I AM 500 YEARS OLD"

        30 END

        RUN

        THIS IS MY FIRST PROGRAM

        MY NAME IS RUMPLESTILTSKIN AND I AM 500 YEARS OLD

        DONE

Note that each PRINT instruction causes the computer to print on a separate line.
If we want a blank line between two printed items (line feed), then we simply type
a line number and the instruction PRINT. This technique will be helpful in creating
labels and headings for more complicated problems as will be seen in later sections
of this chapter.

        The format of a PRINT instruction can be controlled by the use of commas if we
wish to print more than one item on a line. When commas are used between the items
to be printed the teleprinter page is divided into five columns of 15 spaces each.
Thus we can have up to 5 items, no longer than 15 characters each, on a single line.
The items will begin in columns 0, 15, 30, 45, and 60, respectively, with the first
character of each item reserved for the algebraic sign for numeric data. Note that
the first column is actually column number 0.

Example 3.5

        10 PRINT 8, 9, 3, 4, 5

        20 END

        RUN

         8   9   3   4   5

        DONE

This example printed only numbers or numeric information. Our next example prints
alphameric information interspersed with numbers.

Example 3.6

```
1Ø PRINT "THE ANSWER TO PROBLEM NUMBER 23 IS", 88

2Ø END

RUN

THE ANSWER TO PROBLEM NUMBER 23 IS               88

DONE
```

Note in this example that the number 88 is printed starting in column 46 since the
alphameric information requires the first 34 columns, i.e., columns Ø through 33.

The format of a PRINT instruction can also be controlled by the use of the
semicolon between items to be printed. In this case, the items are packed
much more compactly across the teleprinter page. The exact spacing between items
depends upon the nature of the last item printed. For example, if we reprogram the
two examples above with the semicolon instead of the comma, we will get the result
shown below.

Example 3.7

```
1Ø PRINT 8;9;3;4;5

2Ø END

RUN

8     9     3     4     5

DONE
```

Example 3.8

```
1Ø PRINT "THE ANSWER TO PROBLEM NUMBER 23 IS";88

2Ø END

RUN

THE ANSWER TO PROBLEM NUMBER 23 IS 88

DONE
```

Lastly, there are two other characteristics of the PRINT instruction when used
with a comma or a semicolon. First, the use of a comma or a semicolon at the end
of a PRINT line will prevent a line feed, and thus the next item will be printed on
the same line.

Example 3.9

    10 PRINT "THE ANSWER TO PROBLEM NUMBER 23 IS";

    15 PRINT 88

    20 END

When we run the program, the result is the same as above.

    Second, if by chance the items to be printed cannot be accommodated on a single line, then the computer will automatically continue on the following line.

    Finally, the PRINT TAB (N) instruction will cause the teletypewriter to start printing in column N. Again, remember that the first column is column $\emptyset$.

Example 3.10

    10 PRINT TAB (5), "X"

    20 PRINT TAB (10), "Y"

    30 PRINT TAB (15), -7

    40 END

    RUN

        X

             Y

                    -7

    DONE

Note that in this example the letter "X" is first printed in column 5. Then the letter "Y" is printed in column 10 of the next line. Lastly, the number "-7" is printed in column 15 of the following line.

3.4 Arithmetic Operations

    There are five arithmetic operations which can be performed by the computer: addition (+), subtraction (-), multiplication (*), division (/), and exponentiation (↑). The asterisk (*) is used for multiplication rather than the cross (X) to avoid confusion with the letter "X". The up arrow (↑) is used for exponentiation because all characters must be typed on the same line. It should be noted that the new teletypewriters have the symbol ∧ instead of the up arrow (↑) in the upper case "N" position. Also, some systems use the double asterisk (**) instead of the up arrow.

Example 3.11

```
1Ø PRINT 25 + 8; 16 - 4
2Ø PRINT 18 * 4; 15/3; 2 ↑ 3
3Ø END
RUN
 33    12
 72    5    8
DONE
```

Note that in this example line 2Ø could have been written 2*2*2 instead of
2 ↑ 3. The former is done by two multiplications, the latter by using logarithms.
The time to accomplish either of these two operations is approximately the same.
However, as the powers get larger, not only is it more inconvenient to use the "*"
format, but also it is more time consuming.

Lastly, no two symbols (operators) can occur adjacent to one another, e.g.,
8 * + 33 is illegal and results in an error message.

3.5 Priority of Operations

So far we have been evaluating expressions involving two numbers and one
arithmetic operator, so that there could be no possible ambiguity as to the answer.
As the expressions get more complicated, the computer must be instructed as to the
exact order in which the operations are to be performed.

Suppose we wanted to evaluate the expression

$$\frac{2}{5} + 7\left(8 - \frac{9}{3}\right)$$

Consider the following program:

```
1Ø PRINT 2/5+7*8-9/3
2Ø PRINT 2/(5+7)*8-9/3
3Ø PRINT 2/5(5+7*8)-9/3
4Ø END
RUN
 53.4
-1.66667
-2.96721
DONE
```

Note that the three results are different from one another. This indicates
that there is indeed a priority of operations which is used when evaluating

expressions.   This priority is as follows:

1.   Parentheses: starting with the innermost parentheses, left to right.

2.   Exponentiation: in order of appearance, left to right.

3.   Multiplication and Division: in order of appearance, left to right.

4.   Addition and Subtraction: in order of appearance, left to right.

Thus in the above program, we have the following results:

	Line 10	Line 20	Line 30
Original	2/5+7*8-9/3	2/(5+7)*8-9/3	2/(5+7*8)-9/3
Operation 1	.4+7*8-9/3	2/12*8-9/3	2/(5+56)-9/3
2	.4+56-9/3	.166667*8-9/3	2/61-9/3
3	.4+56-3	1.33333-9/3	.0327867-9/3
4	56.4-3	1.33333-3	.0327867-3
5	53.4	-1.66667	-2.96721

Note that none of the above is the equivalent of what we wanted:

$$\frac{2}{5} + 7 \left(8 - \frac{9}{3} \right)$$

which is simply

    2/5 + 7*(8-9/3)

    2/5 + 7*(8-3)

    2/5 + 7 * 5

    .4 + 7 * 5

    .4 + 35

    35.4

We want to reemphasize that algebraic expressions may appear quite different from their BASIC counterparts.

Example 3.12

$$\frac{4 \times 6}{2 \times 3}$$

is not equal to

$$4*6/2*3$$

The former algebraic expression is

$$(4)(6)/(2)(3) = 24/6 = 4$$

The latter BASIC expression is

$$4*6/2*3 = 24/2*3 = 12*3 = 36$$

Lastly, we may use parentheses within sets of parentheses, in which case the expression enclosed by the innermost set is evaluated first, then the next innermost set, etc.

Example 3.13

$(2+(8*9/3)*7)+4$

$= (2+(72/3)*7)+4$

$= (2+24*7)+4$

$= (2+168)+4$

$= 170 + 4$

$= 174$

We wish to point out that the user may find it helpful to use extra parentheses for clarity.

3.6 Number Representation

So far all of the numbers which we have encountered have been either integers such as 8, 5, 9, 4 or decimals with a fractional part such as 1.73, 8.5, 53.4, -1.66667. Although we can enter any decimal number we like, the computer is usually restricted to processing six significant digits*. Therefore, a compromise has to be made in the way numbers are represented. This is accomplished by the use of exponential or scientific notation which is nothing more than a number with up to six significant digits, the decimal point being located in the standard position after the first digit, followed by the letter "E" and a pair of digits which typically ranges from -38 to +38, although this range may vary from computer to computer.

Example 3.14

$$1.20340E+07$$

The part on the left of the E is called the mantissa and tells us only what the digits of the number are. The number to the right of the letter E is called the

*NOTE: The user should check the system he is using for the specific number of significant digits carried.

characteristic and shows the power of ten by which we must multiply the mantissa to
adjust the decimal point to its proper location. Thus in the example above the
number represented is really

$$1.2\emptyset 34\emptyset E+\emptyset 7 = 1.2\emptyset 34 * 1\emptyset^7 = 12,\emptyset 34,\emptyset\emptyset\emptyset$$

An easy way to remember the effect of the "E" format is that when the sign of the
characteristic is positive we move the decimal from the standard position to the
right the number of spaces indicated. On the other hand when the sign is negative,
we move the decimal point to the left as many spaces as indicated. In other words,
the letter "E" stands for the words "times ten to the power".

Example 3.15

     $1.\emptyset 249\emptyset E+\emptyset 7 = 1\emptyset,249,\emptyset\emptyset\emptyset$

     $1.\emptyset 249\emptyset E-\emptyset 7 = .\emptyset\emptyset\emptyset\emptyset\emptyset\emptyset 1\emptyset 249$

     $-1.\emptyset 249\emptyset E+\emptyset 7 = -1\emptyset,249,\emptyset\emptyset\emptyset$

     $-1.\emptyset 249\emptyset E-\emptyset 7 = -.\emptyset\emptyset\emptyset\emptyset\emptyset\emptyset 1\emptyset 249$

NOTE: It is possible to enter data in either the standard decimal format or the
"E" format, but the computer will print out in "E" format whenever necessary.

3.7 Variable Names

     So far we have used the computer like a sophisticated desk calculator. One
important difference between a calculator and a computer is the amount of information that a computer can "remember" or store in its "memory". In order to exploit
the computer's tremendous memory we first have to understand how it stores numbers.
We can think of Figure 3.1 on the following page as representing the computer's
memory, where each box in the table can store one item. Note that each box can be
identified by a "name" consisting of either a single capital letter from A to Z or
a single capital letter followed by a single digit from $\emptyset$ to 9. For example, box A
contains the number 36, box A8 the number 45, box C$\emptyset$ the number 72, and box Z the
number -5.2. All of the other boxes of course are empty. Thus we have the possibility of 26 x 11 = 286 different variable names. We wish to point out that the
computer merely associates a given item with a given variable name and is not concerned with what that number represents. In other words, the item may represent
the population of Clinton, Iowa, the balance in a checkbook, or the time left on a
loan from the bank. Furthermore, although the name is fixed, the item it represents

		Ø	1	2	3	4	5	6	7	8	9
A	36									45	
B											
C		72									
D											
E											
F											
G											
H											
I											
J											
⁞											
Z	-5.2										

Figure 3.1

may vary as a result of programmed computation.   This is the basis of the use of
the words "variable name".

3.8 READ and DATA

The READ and DATA combination of statements gives us a method of getting information into the computer.   For example, suppose we wish to enter the items shown
in Figure 3.1.   The following program will accomplish this:

    1Ø READ A, A8, CØ, Z
    2Ø DATA 36, 45, 72, -5.2
    3Ø END

In this example there is one READ statement together with one DATA statement.   It
is not always necessary to have the same number of READ and DATA statements as long
as there are at least as many data items as there are variable names to be read.
Thus our program above could have been written in either of the following forms:

```
1Ø READ A, A8

2Ø READ CØ

3Ø DATA 36, 45, 72, -5.2

4Ø READ Z

5Ø END
```

or

```
1Ø DATA 36, 45, 72

2Ø READ A, A8, CØ, Z

3Ø DATA -5.2

4Ø END
```

To understand more fully the flexibility of READ-DATA commands, let us look at how the computer goes about interpreting these commands. When we type RUN, the computer doesn't immediately start to compute, but rather starts looking for DATA statements in the program, the search starting from the lowest numbered statements in the program. It gathers all the pieces of data into a library in the same order that they appear in the program. For example, consider the following program:

```
1Ø READ X, Y, Z, A, B

2Ø DATA 33, 44, 55, 66, 68

3Ø DATA 77, 82

4Ø END
```

In this case the library would be as shown in column 1.

33	33←	X 33	X 33	X 33	X 33	X 33
44	44	44←	Y 44	Y 44	Y 44	Y 44
55	55	55	55←	Z 55	Z 55	Z 55
66	66	66	66	66←	A 66	A 66
68	68	68	68	68	68←	B 68
77	77	77	77	77	77	77←
82	82	82	82	82	82	82

When the computer comes to the END statement in line 4Ø it places a pointer or arrow to indicate the piece of data that will be assigned the first name appearing in the first READ statement. Thus, the arrow is at the first number, 33, as shown in column 2. Now, when the computer starts to execute the above program it encounters the first READ statement in line 1Ø. It assigns the value with the arrow, 33, to the name X since X is the first name in the READ statement. The arrow is then advanced to 44 as shown in column 3. The next name encountered in the READ statement happens to be Y, so it is assigned the value with the arrow, 44, and the arrow is again advanced, this time to 55. The next name, Z, is similarly assigned the value of 55, and the arrow is advanced to 66. The procedure is repeated until the value 68 is assigned to B and the arrow is at 77. If there were more READ statements appearing in the program, the process of the name assignment and arrow advancement would continue. If for some reason it happened that there were too many pieces of data for the number of names to be assigned (as in the case in the example above), the computer would assume that we knew what we were doing, ignore the excess data, and follow the remaining instructions in the program. If, however, there were too many names for the data and we asked the computer to READ when in reality there was no new data, the computer would respond with the error message

"OUT OF DATA IN LINE L"

where L is the line number of the corresponding read statement.

Now the procedure for executing a program called for the computer to scan the entire program for DATA statements before it began to execute any instruction. For that reason, DATA statements may come anywhere in the program before the END statement. Some people put them all in the beginning of the program, others put them at the end just before the END statement, and still others in the line following each of the READ statements wherever they may appear. The choice is completely arbitrary as is exemplified by the following programs:

```
1Ø DATA 22,33,44
2Ø READ X,Y,Z
3Ø PRINT X,Y,Z
4Ø END
RUN

 22            33            44

DONE
```

```
1Ø  READ X,Y,Z
2Ø  DATA 22,33,44
3Ø  PRINT X,Y,Z
4Ø  END
RUN
  22              33              44
DONE
```

```
1Ø  DATA 22
2Ø  READ X,Y,Z
3Ø  DATA 33
4Ø  PRINT X,Y,Z
5Ø  DATA 44
6Ø  END
RUN
  22              33          44
DONE
```

Note that each of these programs accomplishes the same thing, namely assigns the numbers 22, 33 and 44 to X, Y and Z respectively, and prints out the current value of X, Y and Z.

Finally, we wish to point out that the READ statement erases the old value of the variables named, and replaces them with the new values specified in the DATA statement(s). On the other hand, the PRINT statement causes the computer to print out the current value of the variables named without erasing them.

3.9 RESTORE

If for some reason we want to restart the reading of data from the beginning we must reset the pointer at the first item in the data library. This is done by using the RESTORE command, as shown in the example on the following page.

Note that in line 2Ø, A is set to 2, B to 5, and C to 4 as in column 2. The RESTORE in line 3Ø then sets the pointer back as in column 3 to the number 2. Finally F and P are set to 2 and 5 respectively, all of which agrees with the result of the PRINT statement. Another example of the RESTORE statement will be seen in Section 4.3.

Data Library

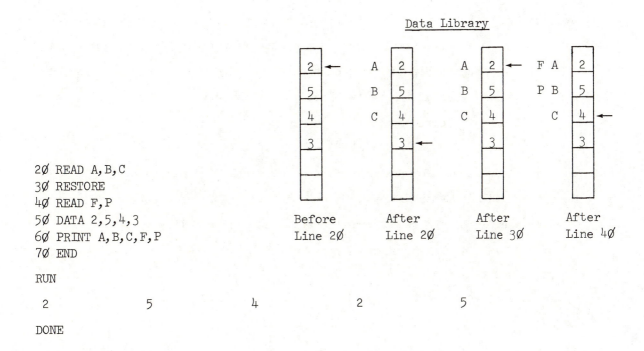

Before   After   After   After
Line 2Ø   Line 2Ø   Line 3Ø   Line 4Ø

```
2Ø READ A,B,C
3Ø RESTORE
4Ø READ F,P
5Ø DATA 2,5,4,3
6Ø PRINT A,B,C,F,P
7Ø END

RUN

 2          5          4          2          5

DONE
```

3.1Ø LET

In Section 3.8 we saw that the READ-DATA combination of statements allows us to enter information into the computer. For example, we saw in that section how we might enter the items in Figure 3.1. An alternate way is by means of a LET statement as shown in the following program:

```
1Ø LET A  = 36
2Ø LET A8 = 45
3Ø LET CØ = 72
4Ø LET Z  = -5.2
5Ø END
```

Just as with the READ-DATA combination this program assigns the number 36 to the name A, 45 to A8, 72 to CØ, and -5.2 to Z.

Although the LET statement contains the "=" sign we must be careful not to interpret it in the usual algebraic sense of "equal to". Rather, it means that the value on the right should be assigned to the name on the left. In this sense the "=" sign simply means "takes the value of", or "is replaced by".

In addition to entering data into the computer, the LET statement allows us to change the values assigned to the variable names. For example, the statement

```
1Ø LET A = A + 5
```

in a program tells the computer to add 5 to the current value of A and then assign
this new value to A.   This example reemphasizes the fact that the "=" sign is not
the algebraic "=", but rather "takes the value of".   Note that the A on the right-
hand side of the "=" sign must have been assigned a value previously in the program
or else we will get the error message

"UNDEFINED VALUE ACCESSED IN LINE 1∅"

So far we have seen two ways of defining a variable; namely, either in a READ
statement or on the left-hand side of an "=" sign in a LET statement.

Another example of the use of the LET statement is as follows:

```
1∅ LET B = 5
2∅ LET C = 2
3∅ LET A = B * C
4∅ PRINT B;C, "THE ANSWER IS";A
5∅ END
RUN
    5    2              THE ANSWER IS 1∅
```

This example assigns the value 5 to B and 2 to C.   Then it calculates the product
of B and C, i.e., 1∅, and assigns the result to A.   Finally, it prints the values
of B, C and the answer A.   Note that the values assigned to both B and C remained
unchanged by the LET in line 3∅.   If there had been a value assigned to A previously, it would have been erased when it was replaced by the product of B and C.

Finally, although not all computer systems require the word LET to be typed
in each LET statement, we will do so.   Furthermore, some computer systems allow
the simultaneous assignment of one value to several names, such as

```
1∅ LET A = B = ∅
```

The user should check his particular system on both of these points.

3.11   INPUT

Still a third way of entering data in addition to the LET and READ-DATA statements is the use of the INPUT statement.   The LET is handy when a one-shot program
is to be run, or when a small number of variables have to be changed to run the
program each time.   The READ-DATA combination is good when many data items are
being changed with each running of the program.   However, sometimes we want to run

the program, see the output, change our mind on a value of a variable, enter that,
and see the results, and so on. Neither the LET nor the READ-DATA statement is
convenient for that.

The INPUT statement allows us to interrupt the program and enter new data before continuing. Consider the following program:

        1∅ INPUT X
        2∅ PRINT "LOOK";X
        3∅ END
        RUN
        ?.76
        LOOK .76

When the computer encountered the INPUT statement in line 1∅, it printed a "?" on
the terminal and waited for us to enter the value that we wished to use for X.   In
this case we entered .76 followed by a carriage return as usual. The computer
then continued with the rest of the program as above.

Just as with the READ statement we may use a single INPUT statement to ask
for a number of values to be entered. For example, the statement

        1∅ INPUT A, B, P3, Z, Z2

will cause the computer to print a "?" on the terminal, at which point we must
enter five items of data. It is important to remember the number of pieces of data
needed and the proper order in which they must be entered. If the order is off,
the name assignments will be off, hence the program may not work as intended. If
the number of pieces of data supplied is not precisely the same as the number of
pieces needed, the computer will tell us that we have a bad format as shown in the
following examples:

Example 3.16

        1∅ INPUT A, B, C
        2∅ PRINT A, B, C
        3∅ END
        RUN
        ?3,4
        ??5

```
      3            4            5
      DONE
```

Example 3.17
```
    1Ø  INPUT A, B, C
    2Ø  PRINT A, B, C
    3Ø  END
    RUN
    ?3, 4, 5, 6
    EXTRA INPUT - WARNING ONLY
      3            4            5
    DONE
```

Note in the first example, when we supplied too few pieces of data, the computer
responded with "??". This is a request for us to supply the missing piece(s) of
data. On the other hand in the second example when we supplied too many pieces of
data, the computer simply notified us and ignored the extra data. One way to avoid
this mismatch of the preceding examples is to precede each INPUT with a PRINT
telling the user exactly what data will be required and in what particular order.
For example, if we add line 5 below to the above programs, we have

```
    5 PRINT "TYPE A, B, AND C"
    RUN
    TYPE A,B, AND C
    ?3,4,5
      3            4            5
    DONE
```

3.12 Flowcharting and Sample Programs

The hardest job for a beginner in writing a computer program is to organize
his thoughts. A useful tool to help him do this is a diagram, known as a flowchart, which is a step-by-step map of the computing process. This map is composed
of special symbols, each representing a BASIC command, together with arrows indicating the flow of the diagram.

In Table 3.1 are the six symbols representing the BASIC commands we have
already used, as well as the interpretation of these symbols. Note that these

symbols are a convention of the authors and are based on the standards suggested by the International Standards Organization (ISO). Below are five examples to illustrate the use of these symbols and their corresponding BASIC program.

Table 3.1

Symbol	Interpretation	BASIC Command
○	Start or end of computing	END
▭	Output	PRINT
▱	Input	READ
▭	Computation	LET
⬯	Input	INPUT
⬡	Other statements	REMARK DATA RESTORE

Example 3.18

Draw the flowchart for the BASIC program of Section 3.1, i.e., print out two numbers.

| Program | Flowchart |

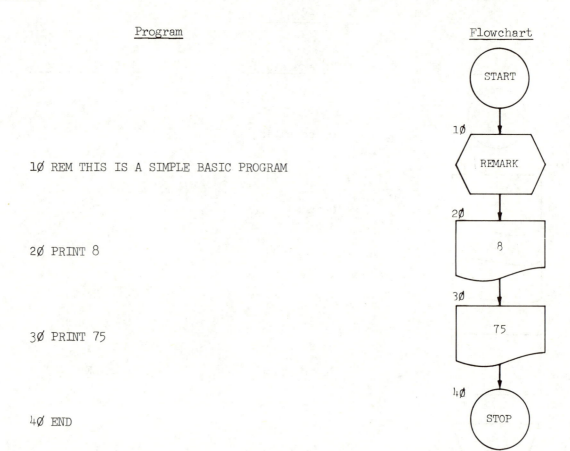

10 REM THIS IS A SIMPLE BASIC PROGRAM

20 PRINT 8

30 PRINT 75

40 END

Note that each line of the program has a corresponding symbol in the flowchart as indicated by the number above the symbol in the flowchart.

Example 3.19

Draw the flowchart for the BASIC program from Section 3.10 which enters two
numbers, computes their product, and prints the result.

Program   Flowchart

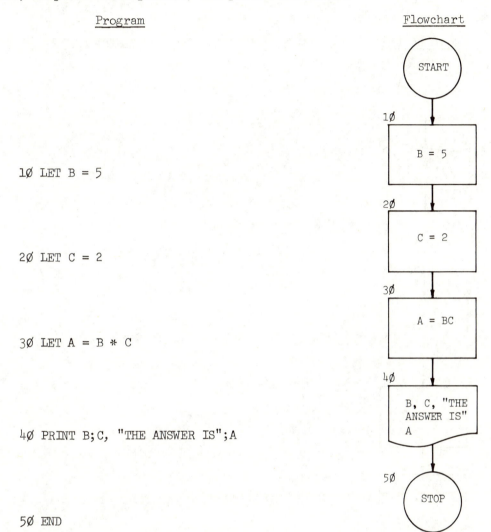

10 LET B = 5

20 LET C = 2

30 LET A = B * C

40 PRINT B;C, "THE ANSWER IS";A

50 END

In the first two examples, we drew the flowchart corresponding to programs which we
had written earlier in this chapter. This is not usually the case, but rather as
we stated in the beginning of this section, the flowchart is drawn first and the
program follows from it, as shown in the following example.

Example 3.20

Draw a flowchart and write the BASIC program to read in three numbers, compute the product of the first two, the quotient of the first and third, and print the numbers and the results.

Flowchart   Program

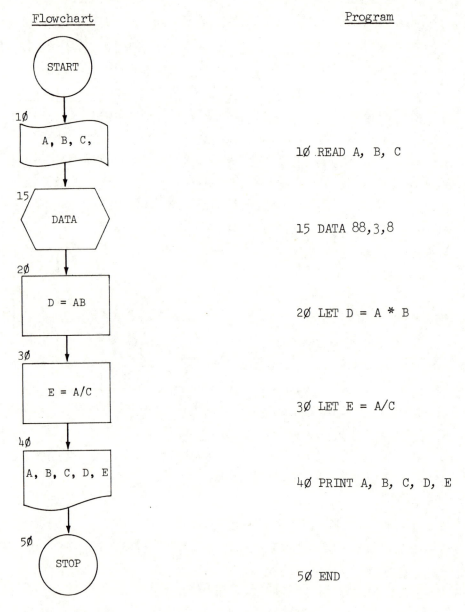

1Ø   1Ø READ A, B, C

15   15 DATA 88,3,8

2Ø   2Ø LET D = A * B

3Ø   3Ø LET E = A/C

4Ø   4Ø PRINT A, B, C, D, E

5Ø   5Ø END

Example 3.21

Draw a flowchart and write the BASIC program to read in five numbers, compute their average, and print the numbers and the averages.

Flowchart   Program

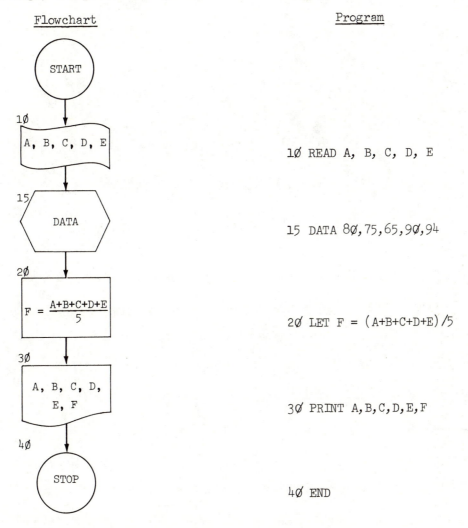

10 READ A, B, C, D, E

15 DATA 80,75,65,90,94

20 LET F = (A+B+C+D+E)/5

30 PRINT A,B,C,D,E,F

40 END

Note that when this program is run the five data items will be printed on one line and the average on the next line. (Recall from Section 3.3 that when the data items are separated by commas in a PRINT statement we can have at most five on a line.)

Example 3.22

Draw a flowchart and write the BASIC program to evaluate the formula $V = P(1 + R)^N$ for several values of P, R, and N to be supplied by the user by way of an INPUT statement. This formula calculates the future value, V, of P dollars invested at R rate of interest compounded annually for N years.

Flowchart   Program

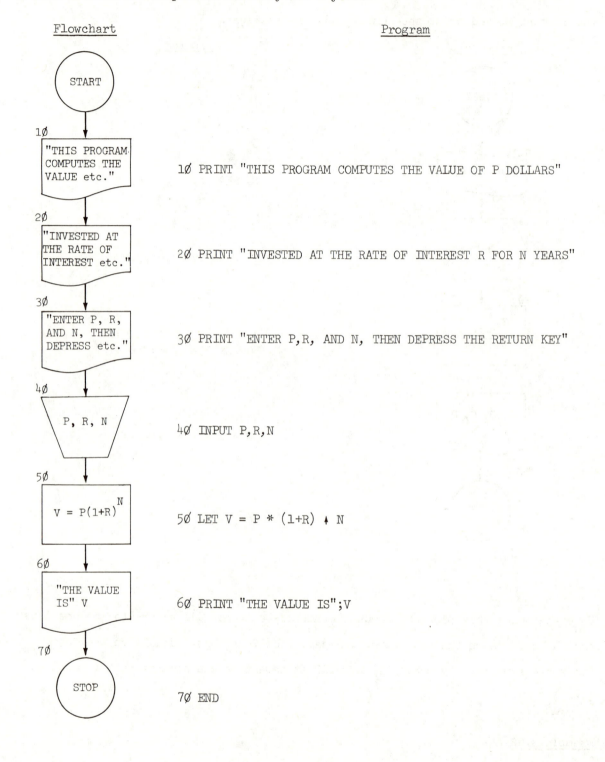

10 PRINT "THIS PROGRAM COMPUTES THE VALUE OF P DOLLARS"

20 PRINT "INVESTED AT THE RATE OF INTEREST R FOR N YEARS"

30 PRINT "ENTER P,R, AND N, THEN DEPRESS THE RETURN KEY"

40 INPUT P,R,N

50 LET V = P * (1+R) ↑ N

60 PRINT "THE VALUE IS";V

70 END

Note that in order to tell the user exactly what data is required for the INPUT statement, we have printed three lines as a heading explaining the program and the order in which the data is to be entered.   Below are the results of running this program with two different sets of values for P, R, and N:

  RUN

  THIS PROGRAM COMPUTES THE VALUE OF P DOLLARS

  INVESTED AT THE RATE OF INTEREST R FOR N YEARS

  ENTER P, R, AND N, THEN DEPRESS THE RETURN KEY

  ?100,.05,11

  THE VALUE IS 171.034

  DONE

  RUN

  THIS PROGRAM COMPUTES THE VALUE OF P DOLLARS

  INVESTED AT THE RATE OF INTEREST R FOR N YEARS

  ENTER P, R, and N, THEN DEPRESS THE RETURN KEY

  ?100,.06,11

  THE VALUE IS 189.83

  DONE

Exercises

1. What will be the result of executing each of the following statements?

    (a) 10 PRINT "8 + 9"

    (b) 10 PRINT 8 + 9

    (c) 10 PRINT "SCHOOL"

    (d) 20 PRINT SCHOOL

    (e) 20 RUN

    (f) LIST

    (g) 10 LIST

2. Write the following expressions in BASIC form and write a program to evaluate those expressions:

	Ordinary Form	BASIC form
1.	$\dfrac{2.3 + 4.57}{4.2}$	$(2.3 + 4.57)/4.2$
2.	$\dfrac{(3.5)^3(4.1)^4}{(4.2)(3+8)}$	
3.	$\dfrac{(4.97)(3.7)}{8.1}$	
4.	$\left(\sqrt{\dfrac{4}{5+18}}\right)^5$	
5.	$2 + \dfrac{3}{1 + 4/3}$	
6.	$\sqrt{3 + \dfrac{4}{(1+7/9)^9}}$	

In Exercises 3 through 8 write a BASIC program which will print the output specified.

3. Print the result of 25 x 57.33.

4. Print your name five times.

5. Print your name and the title of this book on

(a) two lines;

(b) one line.

6. Print the numbers from 5∅ to 6∅ inclusive

(a) all on the same line;

(b) each on a separate line.

7. Print the following in the form given:

1   3   5   7   9   11   13

2   4   6   8   1∅   12

8. Print

BASIC

IS

EASY

TO

LEARN

9. In 1626, the Dutchman Pieter Minuit bought Manhattan Island for $24. What

would the value of that $24 be today if the Indians had invested it at 5%
compounded annually?

Hint: The value at the end of

$$1626 \text{ is } 24(1+.\emptyset5)^1$$

$$1627 \text{ is } 24(1+.\emptyset5)(1.\emptyset5) = 24(1.\emptyset5)^2$$

$$1628 \text{ is } 24(1+.\emptyset5)^2(1.\emptyset5) = 24(1.\emptyset5)^3$$

$$1926 \text{ is } 24(1.\emptyset5)^{3\emptyset1}.$$

10. Complete the following table:

Expression	BASIC Instruction to Evaluate the Expression
(a) $T = \dfrac{a-b}{c-d}$	LET T = (A-B)/(C-D)
(b) $D = a + \dfrac{b}{a+c}$	
(c) $F1 = a + \dfrac{b}{a+\frac{c}{d}}$	
(d) $D = 3x + 4y$	
(e) $D = \dfrac{a}{b} - \dfrac{4c}{3bc}$	
(f) $D = 1 + \dfrac{1}{2+\frac{1}{2+1/2}}$	
(g) $X1 = 3x^2 + 4xy$	
(h) $G3 = 3x^2(xy)^3$	
(i) $Y2 = 4x^2 + \dfrac{3x^2y}{4x^3y^8}$	

11. Supply your own values for variables in the above example and write a program
to evaluate each expression.

12. Suppose that we want to convert distances in feet to distances in meters
(1 meter = 3.28 feet). Write a program to express 13 feet in terms of meters.
Changing only one line at a time have the computer express 18, 2\emptyset, 2\emptyset.4, 13.8
feet in terms of meters.

13. ABC Company manufactures five types of chairs every day.   Let A, B, C, D, and
     E represent the number of chairs manufactured today.   The profit from the dif-
     ferent types of chairs is respectively 5, 6, 8, 11, and 4 dollars per chair.
     Write a program to compute profit for the day if A=2, B=2, C=1, D=8, and E=3.

14. What will be the final values of A, B, and C in the following programs?

     (a)   1∅ LET A = 5   (b)   1∅ LET A = 9
           2∅ LET B = 7   2∅ LET C = A
           3∅ LET B = A + B   3∅ LET B = A + C
           4∅ LET A = B   4∅ END
           5∅ END

15. What will be the result of executing each of the following programs?   If there
     is an error, specify it.

     (a)   1∅ READ P, X, Y   (b)   1∅ READ P, X, Y
           2∅ DATA 4   2∅ DATA 4, 5, 6, 8
           3∅ DATA 5   3∅ END
           4∅ END

16. What values would A, B, and C have as a result of each of the following pro-
     grams?

     (a)   1∅ DATA 1, 1.5   (b)   1∅ DATA 5, 4, 2, 9
           2∅ READ X   2∅ READ A, B, C, D
           3∅ DATA 4, 5, 2, 9   3∅ LET A = A+B
           4∅ READ A, B, C, A1, A2   4∅ END
           5∅ PRINT A1
           6∅ END

17. Write a BASIC program to read 10 numbers and print out the square, reciprocal
     and square root for each number.   (Use READ-DATA).

18. Write a BASIC program which will ask you to supply R, the radius of a circle,
     by an INPUT statement, and print out the area and circumference.   (Area = πR^2
     and Circumference = $2\pi R$; π is approximately 3.1416.)

4

BASIC Commands Continued — Transfer of Control

In this chapter we will discuss the concept of branching and distinguish between unconditional and conditional transfer of control. Then we will discuss repetitive sequences of operations or looping. Finally, we will show how to nest loops within other loops.

4.1 Branching

As we have seen earlier, the computer uses the line numbers in sequential order to control the flow of computations. However, sometimes it is necessary to alter the natural sequence of control and branch or jump to a nonconsecutive line number in order to continue the program. There are two kinds of branches, unconditional and conditional. The unconditional branch must always be taken whenever it is encountered in a program. The conditional branch is taken or not depending on whether a prespecified condition is met. We will discuss each of these branches in the next sections.

4.2 GOTO

The GOTO statement causes an unconditional branch to the line number specified after the word "TO". This branch may be either forward or backward in the program. The GOTO is used to cause the program to simply continue looping through a sequence of operations or loop until a conditional branch diverts the program.

Example 4.1

Repeat Example 3.21 for finding the average of five numbers, this time with three different sets of data.

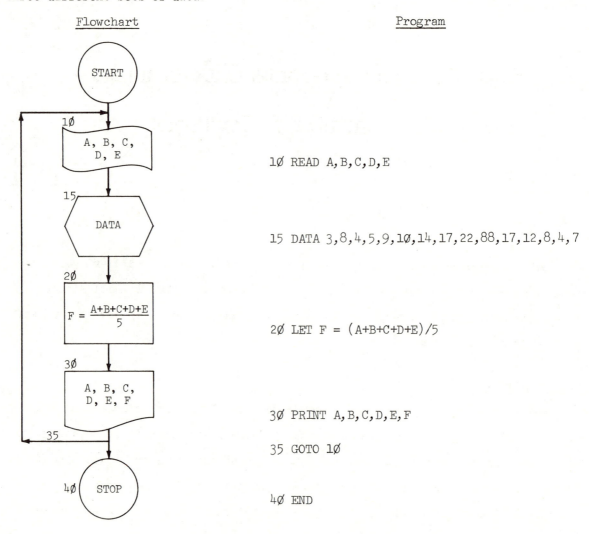

Flowchart   Program

10 READ A,B,C,D,E

15 DATA 3,8,4,5,9,10,14,17,22,88,17,12,8,4,7

20 LET F = (A+B+C+D+E)/5

30 PRINT A,B,C,D,E,F

35 GOTO 10

40 END

Note that the only difference between this flowchart and the previous one is the extra arrow indicating the GOTO in line 35. As we saw before, this program begins by reading in the first five numbers (3,8,4,5,9), computing their average, and printing out the numbers and their average. Now, however, after completing line 30, the computer encounters the GOTO in line 35 which prevents it from going to line 40,

and sends it back to line 1∅. Then the next five numbers (1∅,14,17,22,88) are
read in, averaged, and printed out together with their average. Again line 35
sends the computer back to line 1∅ to repeat the process for the next set of five
numbers (17,12,8,4,7). Once more line 35 sends the computer back to line 1∅, but
this time there is no more data and the computer prints the message OUT OF DATA IN
LINE 1∅ and halts. The results of running this program are shown below.

    RUN

    3   8   4   5   9
    5.8
    1∅   14   17   22   88
    3∅.2
    17   12   8   4   7
    9.6

    OUT OF DATA IN LINE 1∅

Note that this program never reaches the END in line 4∅, and will never execute any
statements with line numbers greater than 35.

Example 4.2

Draw a flowchart and write a BASIC program which will print a table of the integers from 1 on, together with their squares and square roots, under appropriate
headings. See flowchart on following page.

Note that after printing the headings in line 1∅, the computer set N to 1,
printed N, N^2; and $\sqrt{N}$, increased N by 1, and returned to line 3∅ to repeat the
process for N=2, etc. This program would have continued indefinitely had we not
stopped it by depressing the BREAK key after it had printed out the results for the
first ten integers.

4.3 IF-THEN

The IF-THEN statement causes a branch to the line number specified after the
word "THEN", if the condition between the words IF and THEN is satisfied. If the
condition is not met, the next statement in natural order after IF-THEN is executed.

Flowchart

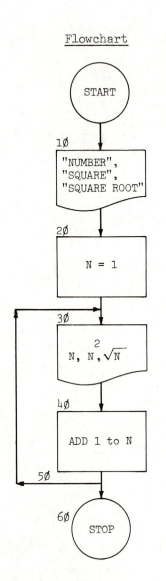

```
1∅ PRINT "NUMBER", "SQUARE", "SQUARE ROOT"

2∅ LET N=1

3∅ PRINT N,N*N,N↑(1/2)

4∅ LET N=N+1

5∅ GOTO 3∅

6∅ END

RUN
```

NUMBER	SQUARE	SQUARE ROOT
1	1	1.
2	4	1.41421
3	9	1.7320∅5
4	16	2.
5	25	2.236∅7
6	36	2.44949
7	49	2.64575
8	64	2.82843
9	81	3.
1∅	1∅∅	3.16228

STOP

In the flowchart we use a lozenge-shaped symbol to represent the IF-THEN statement, as shown below.

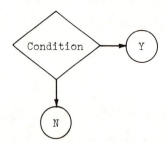

The "Y" and the "N" correspond to the two results of the IF-THEN statement. If there is a transfer, i.e., the condition is true, then the "Y" (Yes) indicates the new flow of the program. If there is no transfer and the next line is executed, then "N" (No) indicates the program flow.

There are six conditions which can be used with the IF-THEN statement. They are:

> Greater than   < = Less than or equal to

< Less than   < > (or #) Not equal to

> = Greater than or equal to   = Equal

Notice the order of symbols in > =, < =, and < > rather than the usual algebraic symbols $\geq$, $\leq$, and $\neq$.

One typical use of the IF-THEN statement is as a test for the end of data. This will avoid the termination of a program with an "OUT OF DATA" message as in Example 4.1 of the previous section, or the use of the BREAK key to stop an "infinite" or non-ending loop as in Example 4.2 of the previous section.

Example 4.3

Use the IF-THEN statement to end the loop in Example 4.1 of the previous section. See flowchart and program on following page.

In this program we first set N to 1 in line 1∅. The variable N is used to count the number of sets of data that we will process. Then we read in the first set of data items, compute the average, and print the results as before in lines 2∅ through 4∅. Next in line 5∅ we increase N by 1 and then in line 6∅ check to see if N is greater than 3; i.e., check to see if we have processed all 3 sets of data. If N is greater than 3, then we branch to line 8∅ and stop. If not, then we proceed to line 7∅ which returns us to line 2∅ to process the next set of data. Note, we used the greater-than test rather than the equal test in line 6∅. This technique is useful because of the way numbers are represented in the computer. That is, the computer operates in binary arithmetic rather than decimal and minor discrepancies can arise. For example, 3.99999 could be the result of an operation and would be interpreted by the user as 4, but would not be equal to 4 when tested by the computer.

Flowchart   Program

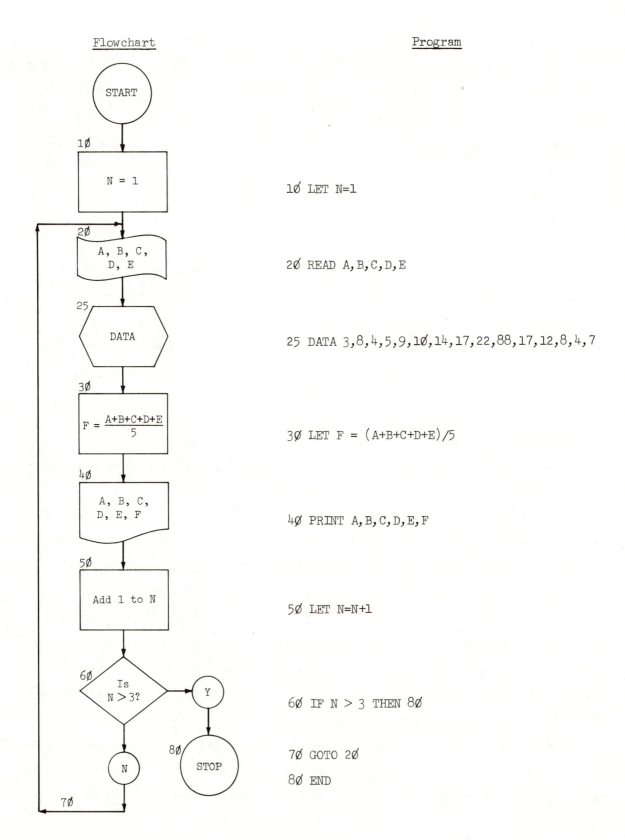

10 LET N=1

20 READ A,B,C,D,E

25 DATA 3,8,4,5,9,10,14,17,22,88,17,12,8,4,7

30 LET F = (A+B+C+D+E)/5

40 PRINT A,B,C,D,E,F

50 LET N=N+1

60 IF N > 3 THEN 80

70 GOTO 20

80 END

Also note that we could eliminate line 7∅ by simply rewriting line 6∅ as

6∅ IF N < 4 THEN 2∅

This is more than a trick since we have thereby saved a line of program instruction, a line in storage, and some time in execution. This may appear trivial but if we were processing several thousand sets of data, the savings in time and money could be appreciable.

A beginner may not always see these shortcuts and savings in his initial attempts at writing programs, but he will become more aware of such techniques through experience.

Another way to test for the end of the data is to place a "dummy" data item or "flag" after the last item of data to be processed. This dummy data should be a number which could not reasonably appear in the data to be processed, e.g., 999999 is a typical dummy. The general procedure is to sequentially read the data and test immediately to see if the item is a legitimate data item or the dummy. In the former case, the data is processed, and in the latter case it is not.

Example 4.4

Draw a flowchart and write a BASIC program which will read an unknown number of data items and find their average. See flowchart and program on following page.

In this program we first set N, which counts the number of data items that we will process, to ∅ in line 1∅. Then in line 2∅ we set S, which will be our running total, to ∅. Next in line 3∅ we read a data item and test immediately in line 4∅ to see if it is the dummy 999999. If it is not, then in line 5∅ we add the latest value of X to the running total S, add 1 to N in line 6∅, and branch back to line 3∅ to read the next data item. On the other hand, if in line 4∅ X is the dummy, then we branch to line 8∅ where we compute the average, print the result and stop.

<u>Flowchart</u>   <u>Program</u>

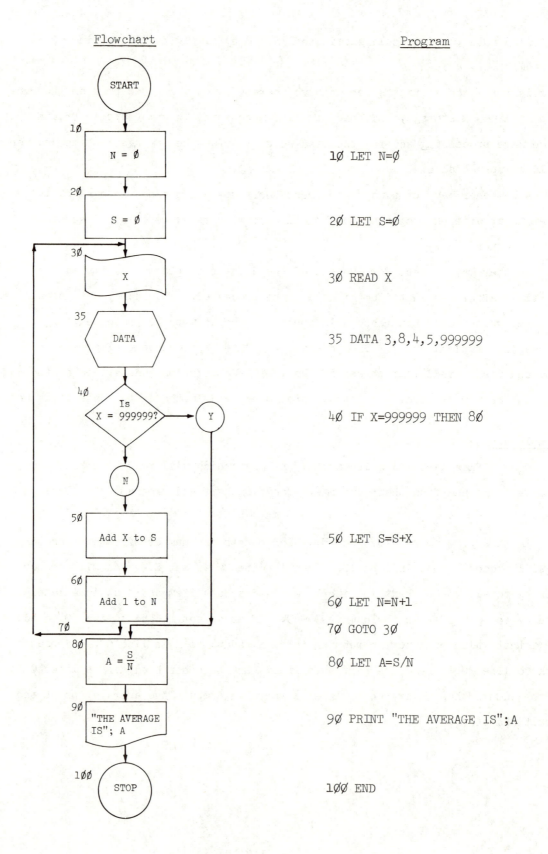

10 LET N=∅

2∅ LET S=∅

3∅ READ X

35 DATA 3,8,4,5,999999

4∅ IF X=999999 THEN 8∅

5∅ LET S=S+X

6∅ LET N=N+1

7∅ GOTO 3∅

8∅ LET A=S/N

9∅ PRINT "THE AVERAGE IS";A

1∅∅ END

In the following table we show the values for N, X, S, and A for each step of the program, where we have underlined the newly defined value in each line.

Loop	Line No.	Value of			
		N	X	S	A
1	10	0			
	20	0		0	
	30	0	3	0	
	40		TEST		
	50	0	3	3	
	60	1	3	3	
2	30	1	8	3	
	40		TEST		
	50	1	8	11	
	60	2	8	11	
3	30	2	4	11	
	40		TEST		
	50	2	4	15	
	60	3	4	15	
4	30	3	5	15	
	40		TEST		
	50	3	5	20	
	60	4	5	20	
5	30	4	999999	20	
	40		TEST		
	80	4	999999	20	5
	90		PRINTS RESULTS		
	100		END		

Note that the reader should always check the results of the loops in a program to make sure that the counter (here we used N) is properly initialized (started). A natural mistake in the above program would be to start in line 10 with N set equal to 1 which would have resulted in a count of five pieces of data when in reality there were only four.

Example 4.5

Refer to the previous example. Now draw a flowchart and write a BASIC program which will compute and print the square of the difference between each data item and the average. To accomplish this, we simply append the following flowchart and program to those in the previous example.

Flowchart   Program

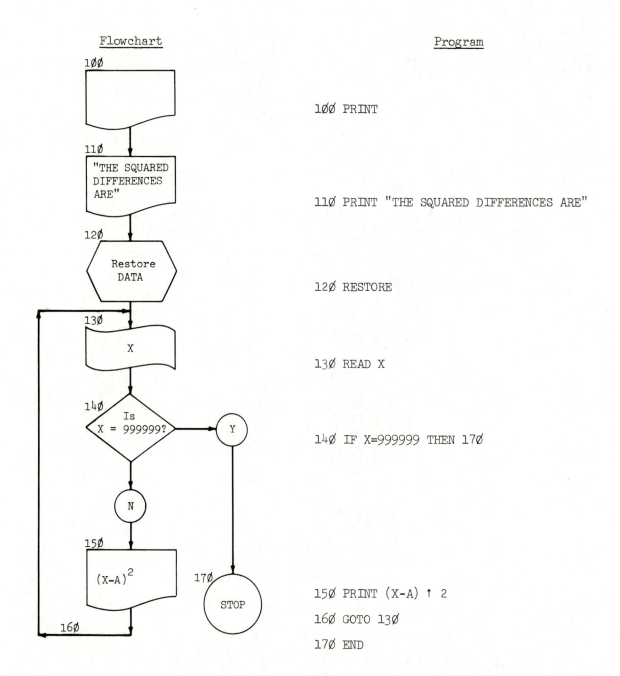

100 PRINT

110 PRINT "THE SQUARED DIFFERENCES ARE"

120 RESTORE

130 READ X

140 IF X=999999 THEN 170

150 PRINT (X-A) ↑ 2
160 GOTO 130
170 END

Note that in this part of the program, line 100 is simply a PRINT statement alone which, as we saw in Section 3.3, will cause the teletypewriter to print a blank line for formatting purposes. Then in line 110, we print a heading for the squared differences, and in line 120 we restore the data, so that, as we saw in Section 3.9, the pointer is reset at the first item in the data library. Next in lines 130 through 160, we compute and print the squared differences until the dummy value of 999999 is reached, at which point the program terminates.

Example 4.6

Modify the flowchart and program of Example 4.1 in this chapter for an unknown number of sets of data.

Flowchart   Program

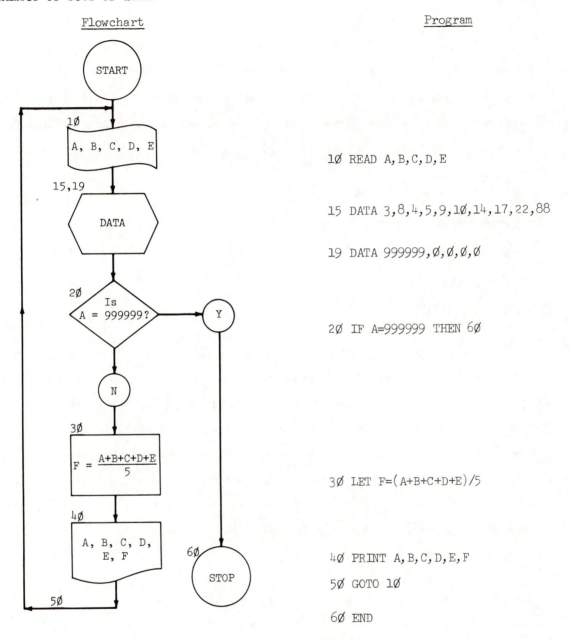

10 READ A,B,C,D,E

15 DATA 3,8,4,5,9,10,14,17,22,88

19 DATA 999999,0,0,0,0

20 IF A=999999 THEN 60

30 LET F=(A+B+C+D+E)/5

40 PRINT A,B,C,D,E,F
50 GOTO 10

60 END

Note that although the test in line 20 is performed on the value of A, we must have some values for B, C, D, and E to go with the dummy 999999, since the READ statement in line 10 is looking for five pieces of data. For this reason in line 19, we have arbitrarily augmented the data with four extra 0's, Also note that by putting the dummy in a separate DATA statement (line 19) the user may use other sets of data later by entering them in line 15 without worrying about reentering the dummy.

4.4   FOR-NEXT

As we discussed in the previous sections, looping is a necessity in many programs.   So far we know how to use a GOTO statement (unconditional branch) or IF-
THEN statement (conditional branch) to carry out a loop in the program.   Now we
would like to discuss a special pair of BASIC statements which are used in place of
the IF-THEN and GOTO to save program lines and for ease in controlling the loop.
This pair is known as FOR-NEXT statements, and is flowcharted in a single symbol
and programmed as shown below.

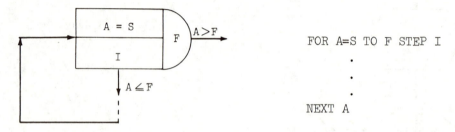

Note that the symbol contains four variables A, S, I, and F.   The first variable A is called the index and is primarily used as a counter to control the number
of times the loop is executed.   The second variable S is called the starting value
and is the initial value assigned to the index A.   The third variable I is called
the increment and is the value by which the index A is increased each time the loop
is executed.   The fourth variable F is called the final value and is used to terminate the looping process.   This termination occurs when the index A is incremented
to a value strictly greater than F.

Example 4.7

Consider the flowchart and program on the following page.

When the computer encounters the FOR statement in line 1∅ it sets the index A
to its initial value of 1.   It proceeds to execute line 2∅, as can be seen in the
printout, and then encounters the NEXT statement in line 3∅.   At this point the increment I=1, specified after the word STEP in line 1∅, is added to the index A and
the new index value A=2 is compared to the final value of F=3, specified after the
word TO in line 1∅.   If the new index value of A is not greater than F the computer

will return to the first statement after the FOR statement, i.e., line 2∅. If, however, A is greater than F the computer proceeds to the first statement after the NEXT, i.e., line 4∅. In this case, because A=2 and F=3, A is not greater than F and hence the computer returns to line 2∅. This process will be repeated until A is 4 and exceeds F=3, as shown in the table below.*

<u>Flowchart</u>
<u>Program</u>

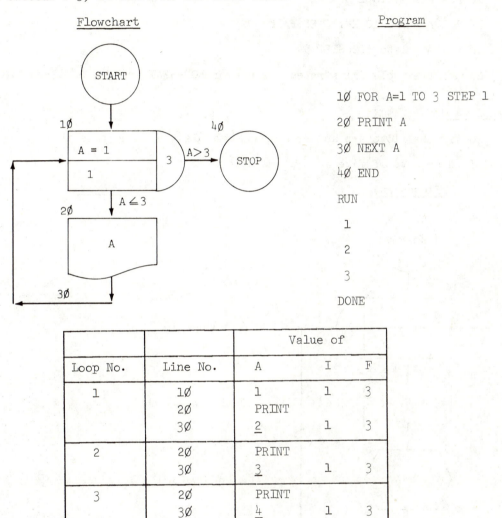

```
1∅ FOR A=1 TO 3 STEP 1
2∅ PRINT A
3∅ NEXT A
4∅ END
RUN
   1
   2
   3
DONE
```

Loop No.	Line No.	Value of		
		A	I	F
1	1∅	1	1	3
	2∅	PRINT		
	3∅	2	1	3
2	2∅	PRINT		
	3∅	3	1	3
3	2∅	PRINT		
	3∅	4	1	3
	4∅	END		

Note, whenever the step is 1, as here, the word STEP and the increment 1 may be omitted in the FOR statement; i.e., the line 1∅ above could have been written as

    1∅ FOR A = 1 TO 3

*NOTE: The user should check the system he is using since some systems leave the value of A at the largest value that does <u>NOT</u> exceed F.

Also note that S, F, and I need not be integers.   Furthermore, I may be negative
for counting backwards, but, in this case, the reader must be sure that F is less
than S or else the FOR-NEXT loop is merely bypassed.   Lastly, the values of S, F,
and I may be algebraic expressions.   Some examples are:

    10 FOR A = 99 TO 1 STEP-2

    10 FOR A = B*C TO X-Y STEP P/10

    10 for A = 1 to 100 STEP 5

Our next example illustrates the use of FOR-NEXT instead of IF-THEN and GOTO.

Example 4.8

    Modify the flowchart and program of Example 4.3 to compute the average of
each of three sets of five data items apiece.

Flowchart   Program

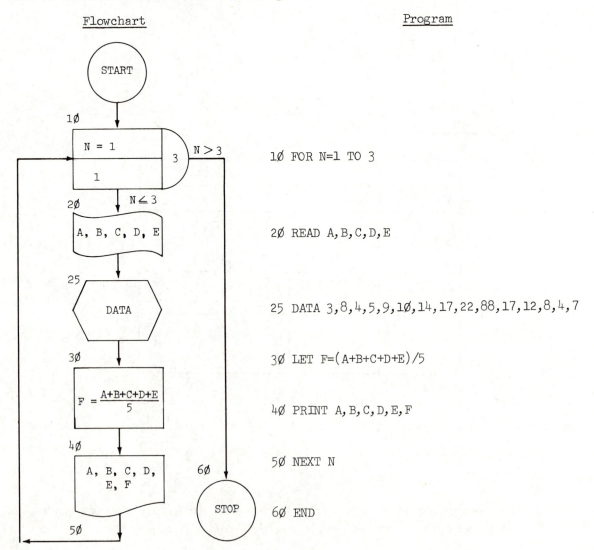

10 FOR N=1 TO 3

20 READ A,B,C,D,E

25 DATA 3,8,4,5,9,10,14,17,22,88,17,12,8,4,7

30 LET F=(A+B+C+D+E)/5

40 PRINT A,B,C,D,E,F

50 NEXT N

60 END

Note that in line 1∅ we have omitted the STEP of 1, and that this program has two fewer program lines than in Example 4.3.

Our next example illustrates the use of the FOR-NEXT in conjunction with the IF-THEN.

Example 4.9

Suppose we want to find the total number of employees paid and the total wages paid in a given week by a company which employs 5∅∅ workers, not all of whom may be paid in that week. In this situation we can set up a loop for reading the pay information for a maximum of 5∅∅ workers, but terminate the looping as soon as we encounter the last piece of data, as identified by the dummy 999999. The flowchart and program on the following page accomplish this.

Notice in line 2∅ the final value of the loop is 5∅1 rather than 5∅∅. The purpose of this is to detect the possibility of too many data items, indicating an error in data preparation. Of course if all employees were to be paid there should be 5∅1 data items including the dummy, and hence the loop should terminate itself before N is greater than 5∅1.

Also note that whenever the loop terminates by encountering the dummy, line 4∅, the count N will include the dummy and hence we subtract 1 from N in line 9∅ before printing the results.

4.5 Nested FOR-NEXT

Often the user may want to vary two or more indices simultaneously within a given program. This can be accomplished by "nesting" one FOR-NEXT loop inside another with the only precaution being that the FOR-NEXT inner loop must be completely contained within the FOR-NEXT outer loop. One way to check that the loops are properly nested is to draw connecting arrows between the corresponding FOR and NEXT statements, making sure that the arrows do not cross one another.*

*NOTE: The user should check the system he is using to see what the limit is for the maximum number of nested loops.

Flowchart   Program

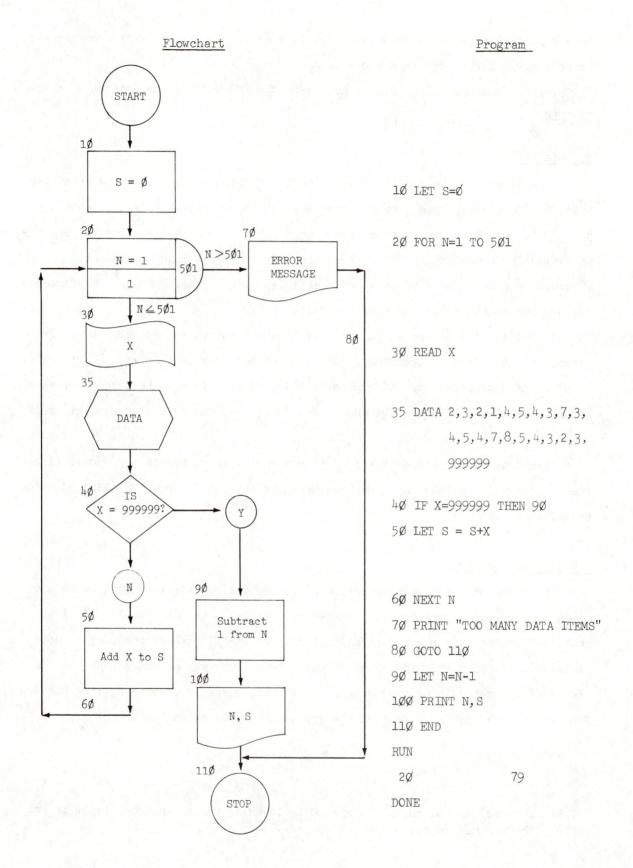

10 LET S=0

20 FOR N=1 TO 501

30 READ X

35 DATA 2,3,2,1,4,5,4,3,7,3,
      4,5,4,7,8,5,4,3,2,3,
      999999

40 IF X=999999 THEN 90
50 LET S = S+X

60 NEXT N
70 PRINT "TOO MANY DATA ITEMS"
80 GOTO 110
90 LET N=N-1
100 PRINT N,S
110 END
RUN
 20   79
DONE

Below are shown both the correct and incorrect ways to nest loops:

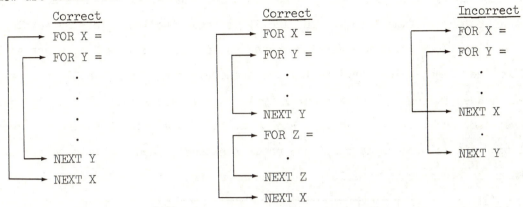

Correct	Correct	Incorrect
FOR X =	FOR X =	FOR X =
FOR Y =	FOR Y =	FOR Y =
.	.	.
.	.	.
.	NEXT Y	NEXT X
.	FOR Z =	.
NEXT Y	.	NEXT Y
NEXT X	NEXT Z	
	NEXT X	

Example 4.10

Draw a flowchart and write a BASIC program which prints a multiplication table for the integers from 1 to 5 inclusively.

Flowchart   Program

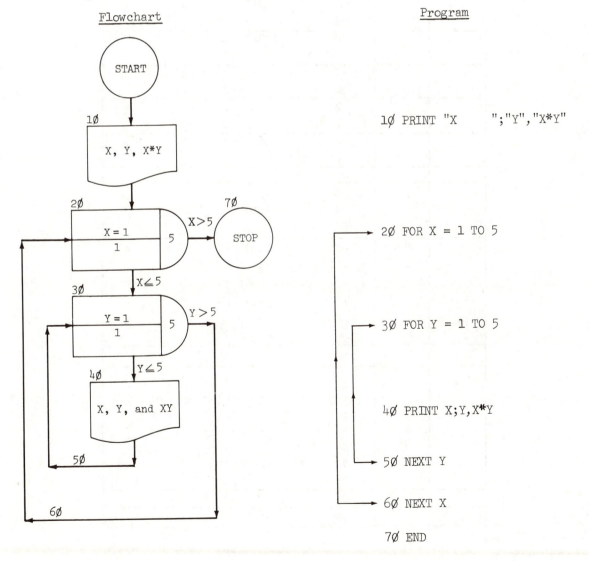

```
1Ø PRINT "X        ";"Y","X*Y"



2Ø FOR X = 1 TO 5



3Ø FOR Y = 1 TO 5



4Ø PRINT X;Y,X*Y



5Ø NEXT Y


6Ø NEXT X

7Ø END
```

In line 1∅ we first print out the headings for the table. In line 2∅ we begin the outer loop by setting X to 1. In line 3∅ we begin the inner loop by setting Y to 1. In line 4∅ we print the entries in the table. Now holding X constant at 1, we follow the normal rules for FOR-NEXT loops incrementing Y by 1 until it exceeds 5. When Y does exceed 5, we increment X by 1 as usual and return to line 3∅ to redo the inner loop on Y; i.e., set Y to 1 and print the table while X is 2 and Y goes from 1 to 5. This process is continued until X exceeds 5 and the program stops.

Below is a summary of the looping process and the associated results.

X LOOP	Y LOOP	LINE NO.		VALUE OF X	Y	X*Y
1		1∅	PRINT	HEADINGS		
		2∅		1		
	1	3∅		1	1	
		4∅	PRINT	1	1	1
		5∅		1	2	
	2	4∅	PRINT	1	2	2
		5∅		1	3	
	3	4∅	PRINT	1	3	3
		5∅		1	4	
	4	4∅	PRINT	1	4	4
		5∅		1	5	
	5	4∅	PRINT	1	5	5
		5∅		1	6	
		6∅		2	6	
2	1	3∅		2	1	
		4∅	PRINT	2	1	2
		5∅		2	2	
	2	4∅	PRINT	2	2	4
		. . . .				
		. . . .				
		. . . .				
	5	4∅	PRINT	2	5	1∅
		5∅		2	6	
		6∅		3	6	
3	1	3∅		3	1	
		. . . .				
		. . . .				
		. . . .				
5	5	3∅		5	5	
		4∅	PRINT	5	5	25
		5∅		5	6	
		6∅		6	6	
		7∅	END			

X	Y	X*Y
1	1	1
1	2	2
1	3	3
1	4	4
1	5	5
2	1	2
2	2	4
2	3	6
2	4	8
2	5	1∅
3	1	3
3	2	6
3	3	9
3	4	12
3	5	15
4	1	4
4	2	8
4	3	12
4	4	16
4	5	2∅
5	1	5
5	2	1∅
5	3	15
5	4	2∅
5	5	25

4.6 In Retrospect

Before proceeding to more BASIC instructions, we would like to pause and re-
flect on what we have done so far. We have introduced the user to a fundamental
set of BASIC instructions and showed, through many examples, how these instructions
can be used to solve many typical problems via the computer.

We would like to point out to the user that the hardest thing for the beginner
to do is to get his program organized. As we saw, a helpful tool is the flowchart,
and we highly recommend its use at all times. Also, the following checklist may
help the user to define for himself the proper input, internal processing, and
output.

1. What are the data required?

2. What are the calculations to be performed?

3. Are there points at which the calculations can be checked; what are the

appropriate courses of action?

4. Are there points at which the data can be checked for credibility?

5. What is the form of the output?

After thoroughly reviewing this check list, the user should now draw a rough flowchart of the computation process and review it for accuracy. If it looks right, he should redraw the flowchart carefully making the program line number annotations. Now the user is ready to program and test the program. Notice the word TEST; it is probably the single most important word the user has met in this book. He should never accept a number or conclusion from a computer as meaningful unless it has been tested for accuracy. The computer will rarely miscalculate what the user tells it to do; however, the catch lies in understanding what it is that the user has really told the computer. What he asks it to do may be entirely different from what he wants it to do.

Exercises

Draw a flowchart and write a BASIC program for each of the following.

1. Input two integers. Calculate the squares of all integers between the first and second number inclusively, printing the numbers and their corresponding squares as you go. After printing the last number, print the sum of squares and the message, "I AM GREAT".

2. Calculate the wages of 1Ø workers each of whom earns $3.75 an hour up to 4Ø hours and $5.5Ø an hour for all hours over that time. Use the READ statement to enter the weekly total of hours for each person. End the program when you encounter -1 for the number of hours. Print "FINISHED ALL MY LABORS". For data, assume the following hours: 41, 32, 45, 52, 38, 39, 47, 46, 18, 1Ø.

3. Type out all the integers from 99 to 115 inclusively.

4. Read in 5Ø numbers, find and print their sum.

5. Print all the odd integers from 11 to 99 inclusively.

6. Print out a table of squares, square roots, cubes, and cube roots for all numbers from 1 to 25. Use the following title for your printout

    NUMBER   SQUARE   SQR   CUBE   CUBE ROOT

7. Do as in 5 above, but only for even numbers from 1 to 3Ø inclusively.

8. The quadratic equation $Ax^2 + Bx + C = \emptyset$ has two roots:

$$\frac{-B + \sqrt{B^2 - 4AC}}{2A}$$

and

$$\frac{-B - \sqrt{B^2 - 4AC}}{2A}$$

If $B^2-4AC<0$, then the roots are imaginary. Read in A, B and C and check whether $B^2-4AC<0$. If $B^2-4AC<0$, print out the message "ROOTS NOT REAL". If $B^2-4AC=0$ print the double root. Otherwise, print the two roots.

9.   Read three numbers, arrange and print them in ascending order. Use 3,4,2; 4,9,1; 14,-5,-2 as data.

10. Read three numbers, arrange and print them in descending order using the above data.

11. Read a set of 10 positive numbers followed by -1, a dummy number. After all data are read in, print out the smallest number (excluding -1).

12. Print the following sequence for n = 1 to 25.

n	sequence
1	1
2	3
3	6
4	10
5	15
6	21
7	28
.	.
.	.

(Hint: the nth term of the sequence is given by the previous term plus n.)

13. Print the following two sequences S1 and S2 as shown:

Number	S1	S2
1	1	1
2	1.5	4
3	1.833	9
4	2.083	16

(Hint: the nth term of S1 is given by the previous term + 1/n).

14. Accept two numbers using an INPUT statement. If the square of the first number is larger than the second number go on to read another pair of numbers;

otherwise print "LARGER" and then read another pair of numbers. Terminate when the first number read in is $1.0E+10$.

15. Read in at most 25 numbers, and have the computer stop reading when the number read is $1.0E+10$. Have the computer print out the number of data items actually read in.

16. Read and count a set of numbers. Stop reading when number is -99. Print the count and "ODD" if the count is odd; otherwise print "EVEN".

17. The present value P of V dollars at the end of N years is $P=V(1+R)^{-N}$ where R is the annual rate of interest on one dollar. Input V, R, and N and print out a table of present values for years 1, 2, 3, ... , N.

18. Modify the above program to print out P1, the cumulative sum of the present values, and stop when $P1 \geq 1.5P$.

19. The method of straight line depreciation means that a capital investment declines in value at a steady rate over the life of the investment. The method applied to a machine worth $5000 which lasts five years is shown below.

Year End	Value Straight Line
1	$4000
2	$3000
3	$2000
4	$1000
5	$ 0

Input the value of an investment and its length of life. Compute and print the value of the investment at the end of each year of its life.

20. Fibonacci was an Italian mathematician of the thirteenth century. He discovered a series of numbers named after him. It begins with 1 and 1. Hereafter, the next Fibonacci number is the sum of the last two numbers; i.e., 1, 1, 2, 3, 5, 8, 13, 21, 34,.... Compute and print all Fibonacci numbers less than 10,000.

5

Specialized BASIC Commands — Subscripts, Matrices, Functions, Strings, and Subroutines

In this chapter, we introduce arrays: first singly subscripted variables often called vectors, or lists, and then doubly subscripted variables often called matrices or tables. Next we discuss the preprogrammed matrix operations available in BASIC. Then we introduce the concepts of library functions and user defined functions. Finally, we discuss strings and subroutines.

5.1 Singly Subscripted Variables

So far we have had to use a separate name for each variable in a computation. Often the number of variable names needed in a particular program becomes prohibitive. A way to overcome this problem is to use singly subscripted variables, i.e., a single capital letter from A to Z identifying the list, followed by a positive integer in parentheses (the subscript) which identifies the location of the data item in the array.

Example 5.1

Consider the following lists of numbers:

L	R
1	5
7	7
3	13
8	18
9	23

Note that the list on the left is identified by the letter L, and the one on the
right by the letter R. Note further that we can identify a particular data item
by the row in which it is written. Thus R(3) is the third entry in the R list
or the number 13; R(5) is 23; L(2) is 7; etc.

The user should be aware that the acceptable names for arrays are limited to
single letters. This should be distinguished from acceptable names for variables
which, as we saw in Section 3.7, consisted of either a single letter or a single
letter followed by a single digit from 1 to 9. Thus A, C, and Z are acceptable
names for both arrays and variables. However, A2 refers to an individual variable
while A(2) refers to the second value in the A list.

To enter the data items into the computer for the L array in Example 5.1 above,
we can as usual use either the LET, READ, or INPUT statements. Thus any of the
following three programs would accomplish this task:

```
1Ø LET L(1) = 1
2Ø LET L(2) = 7
3Ø LET L(3) = 3
4Ø LET L(4) = 8
5Ø LET L(5) = 9
6Ø END
```

```
1Ø READ L(1),L(2),L(3),L(4),L(5)
2Ø DATA 1,7,3,8,9
3Ø END
```

```
1Ø FOR I = 1 TO 5
2Ø INPUT L(I)
3Ø NEXT I
4Ø END
```

Note that in the program using the INPUT statement, the computer will ask five
times for the user to supply the necessary data items.

The user must be cautioned that if he wishes to use a single letter to represent
a list rather than a variable, he must alert the computer to reserve enough space
to store the numbers in the list. This is done by means of a DIM (dimension)
statement prior to the first usage of the array. For example,

10 DIM A(25)

tells the computer to reserve space for 25 data items under the array named A.
(The user should check the system he is using since some do not require a DIM
statement where there are 10 or fewer data items.)

5.2 Doubly Subscripted Variables--Matrices

As we have just seen, we can use a capital letter with a single subscript to
identify an array or list of values. Next, we would like to discuss an extension
of this to matrices or tables. In this case we use a capital letter with two subscripts, the first subscript identifying the row in which the data item occurs, and
the second identifying the column.

Example 5.2

Consider the following table:

Table B

Rows \ Columns	1	2	3	4
1	1	7	9	8
2	7	8	4	3
3	1	7	18	15

Note that this table has 3 rows and 4 columns so that there are 12 data items. In
particular $B(3,4)$ refers to the data item in the third row and fourth column of the
B table; namely 15. Similarly $B(2,3)$ is 4; $B(3,2)$ is 7, etc.

As with a list we can enter the data items into the B table above by means of
LET, READ, or INPUT statements. One way of doing this is by way of a nested FOR-
NEXT loop as shown below.

Example 5.3

```
10 FOR R = 1 TO 3
20 FOR C = 1 TO 4
30 READ B(R,C)
35 DATA 1,7,9,8,7,8,4,3,1,7,18,15
40 NEXT C
50 NEXT R
60 END
```

Note that the C (column) loop is nested within the R (row) loop so that the data
items will be entered row by row, i.e., as seen in the following table:

R Loop	C Loop	Line No.		R	C	Value of B(R,C)
1		10		1		
	1	20		1	1	
		30	READ	1	1	1
		40		1	2	
	2	30	READ	1	2	7
		40		1	3	
	3	30	READ	1	3	9
		40		1	4	
	4	30	READ	1	4	8
		40		1	5	
		50		2	5	
2	1	20		2	1	
		30	READ	2	1	7
		40		2	2	
	2	30	READ	2	2	8
		. . . .				
		. . . .				
3	4	30	READ	3	4	15
		40		3	5	
		50		4	5	
		60	END			

Just as with a list the user must be cautioned in the use of a matrix to alert the computer to reserve enough space to store the data items. This is done once again by the DIM statement with two integer entries, the first specifying the number of rows and the second the number of columns. For example,

    10 DIM A(25,25)

tells the computer to reserve space for 25 times 25 or 625 data items under the matrix named A. (The user should check the system he is using since some do not require a DIM statement for a matrix which is 10 by 10 or smaller.)

Example 5.4

Draw a flowchart and write a BASIC program which will read in the two matrices below, add them together (i.e., add their corresponding elements), and print the result.

$$A = \begin{pmatrix} 1 & 5 & 7 \\ 2 & 4 & 6 \end{pmatrix} \qquad B = \begin{pmatrix} 7 & 12 & 4 \\ 8 & 9 & 2 \end{pmatrix}$$

Specialized BASIC Commands   69

Flowchart

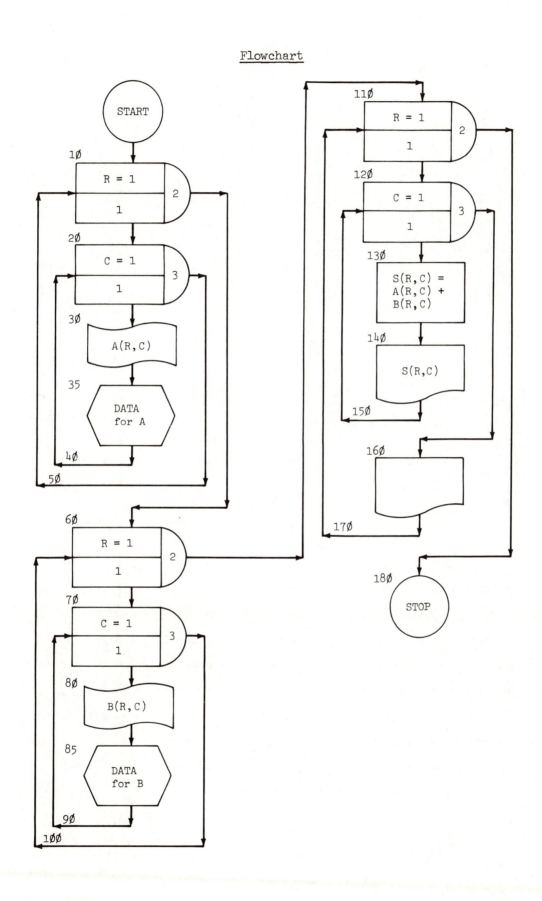

<u>Program</u>

```
10   FOR R=1 TO 2
20   FOR C=1 TO 3
30   READ A(R,C)
35   DATA 1,5,7,2,4,6
40   NEXT C
50   NEXT R
60   FOR R=1 TO 2
70   FOR C=1 TO 3
80   READ B(R,C)
85   DATA 7,12,4,8,9,2
90   NEXT C
100  NEXT R
110  FOR R=1 TO 2
120  FOR C=1 TO 3
130  LET S(R,C)=A(R,C)+B(R,C)
140  PRINT S(R,C);
150  NEXT C
160  PRINT
170  NEXT R
180  END
```

Note that we have presented the flowchart and the program on separate pages because of their length. Note also that in the program, statements 10 through 50 read in matrix A, and 60 through 100 read in matrix B. Then statements 110 through 170 compute and print each of the elements of the sum matrix S. We wish to point out that the semi-colon used at the end of the PRINT statement in line 140 prevents the normal line feed and carriage return, and the PRINT statement in line 160 will cause the desired line feed when a complete row of the matrix S has been printed.

The results of running this program are as follows:

```
RUN
  8      17     11
 10      13      8
DONE
```

5.3 Matrix Operations

We have seen in Example 5.4 of the previous section how to add two matrices
through the use of doubly subscripted variables and nested loops. BASIC provides
a set of instructions, called MAT instructions, which allow us to manipulate
matrices more easily. In particular, the program for matrix addition could have
been written as follows:

```
1Ø DIM A(2,3),B(2,3),S(2,3)
2Ø MAT READ A
25 DATA 1,5,7,2,4,6
3Ø MAT READ B
35 DATA 7,12,4,8,9,2
4Ø MAT S = A+B
5Ø MAT PRINT S
6Ø END
```

Note that in this program we first dimensioned the matrices A, B, and S in
line 1Ø. DIM statements must always be used for each matrix appearing in any MAT
statement in the program. Such DIM statements not only reserve space for the data
items but also specify the exact size of each matrix being operated upon. For this
reason, the user must be extremely careful in the dimensioning of matrices. Next
in line 2Ø we read in the matrix A row-by-row using the MAT READ instruction. Note
that this single instruction replaces lines 1Ø, 2Ø, 3Ø, 4Ø, and 5Ø of Example 5.4 in
the previous section. Similarly the MAT READ B instruction in line 3Ø of our program above replaces lines 6Ø, 7Ø, 8Ø, 9Ø, and 1ØØ of our previous program.
Finally, lines 4Ø and 5Ø of our program above add the matrices together and print
the result as was done in lines 11Ø through 17Ø of our previous program.

We have just seen an example of how the MAT instructions can simplify the
writing of a program. The following table contains the most commonly used MAT
instructions and their corresponding interpretations.

MAT Instructions

Instruction	Interpretation
MAT A = ZER	Sets each value in the A matrix to zero.
MAT A = CON	Sets each value in the A matrix to one.
MAT INPUT A	Allows user to input values in the A matrix row by row, within the limits set by DIM statement.
MAT READ A	Reads values in A matrix row by row.
MAT PRINT A	Prints values of A matrix row by row.
MAT A = IDN	Sets up identity matrix.
MAT C = A + B	Adds matrices A and B and stores in C.
MAT C = A - B	Subtracts matrices A and B and stores in C.
MAT A = B * C	Multiplies matrices B and C.
MAT A = INV(B)	Finds the inverse of B.
MAT A = TRN(B)	Transposes matrix B.
MAT A = (K) * B	Multiplies the matrix B by the scalar K.

5.4 Library Functions

Before discussing the use of library functions, we wish to review the general concept of a function! As the user may recall, a function is a rule which associates with a given object, one and only one other object. The given object is often called the argument of the function and the other object is often called the value of the function. For example, consider the function defined by the phrase "is the mate of". Then if we have a pile of left shoes and another pile of right shoes, we can associate with each of the lefts its mate on the right.

In this book, we are concerned solely with numerically valued functions, i.e., rules which associate with a given number, one and only one other number. For example, consider the absolute value function which is ordinarily written as

$f(x) = |x|$

where x is the argument and $f(x)$ is the value. As the user may recall, this function associates with each number the magnitude of that number so that

$f(2) = |2| = 2,$

$f(-2) = |-2| = 2,$

$f(\emptyset) = |\emptyset| = \emptyset,$

etc.

The library of the computer contains many preprogrammed functions, one of which is the absolute value function we just discussed. This function can be used as follows:

    1Ø LET Y = ABS(X)

In this case, the computer would take the absolute value of the number stored in X and place the result in Y.

The table below contains the most commonly used library functions and their corresponding interpretations.

Function	Interpretation
ABS(X)	Finds the absolute value of the argument.
SIN(X)	Finds the sine of the argument. Note: The argument is assumed to be expressed in radians.
COS(X)	Finds the cosine of the argument. (See note above).
TAN(X)	Finds the tangent of the argument. (See note above).
LOG(X)	Finds the logarithm of the argument. Note: The base used is not the common base of 1Ø, but rather the mathematical constant e, which is about 2.7182.
EXP(X)	Finds the value of the constant e raised to the power of the argument.
RND(X)	Finds a random number between Ø.Ø and Ø.999999, approximately. See further description below.
INT(X)	Finds the next integer equal to or less than the argument. See further description below.
SQR(X)	Finds the non-negative square root of the argument.

In all of the above cases, we referred to the argument X and not to a specific number. Again, BASIC is versatile, and can handle either a number or an expression in BASIC as the argument of a function. For example, consider

    1Ø LET A = SIN(67/C-2Ø)

where C had been previously defined in the program. The computer would find the value of 67/C, subtract 2Ø, find the sine of the result of the whole expression, and set A to that value.

There are two functions in the above list which we would like to explain further. First, the RND(X) will in general give the user a different random number each time that it is used. Further, the argument X of this function serves no purpose other than to conform to the standard BASIC definition of a function; i.e., the X is a dummy variable which may either be a previously defined variable or a number, and the random numbers will remain unchanged regardless of the value of the argument.*

Second, the INT(X) will always round the value of X down to the largest integer less than or equal to the argument X.** This is illustrated by the entries in the table below.

X	INT(X)
115	115
33.8	33
10.73	10
-1	-1
-1.5	-2

Note that in the case of a negative number such as -1.5, the rule still holds; i.e., the value of INT(-1.5) is -2 not -1.

Our next example illustrates the use of two library functions, RND and INT, in the same program.

Example 5.5

Draw a flowchart and write a BASIC program which will print 1Ø random integers, each between 1 and 5 inclusively. See flowchart and program on following page.

Note that the random integer between 1 and 5 is obtained in line 2Ø as follows. First, we take the RND(Ø) which generates a random number between Ø.Ø and Ø.999999. Next, we multiply this random number by 5 to obtain a random number between Ø.Ø and 4.99999. Then, we add 1 to this to generate a random number between 1.Ø and 5.99999. Finally, we take the integer value of this number using INT to generate the random integer between 1 and 5 inclusively. The results of running

---

*NOTE: The user should check the system he is using since in some systems the argument is not a dummy variable.

**NOTE: The user should check the system he is using for the specific rounding technique used by INT(X).

Flowchart   Program

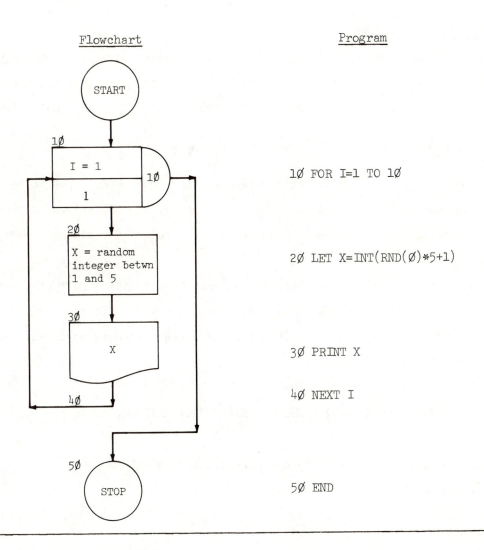

10 FOR I=1 TO 10

20 LET X=INT(RND(∅)*5+1)

30 PRINT X

40 NEXT I

50 END

this program are as follows:

    RUN
      5
      4
      5
      4
      4
      3
      5
      1
      5
      1
    DONE

5.5 User Defined Functions

In the previous section, we saw that in the computer there are a number of preprogrammed or library functions. In addition to these functions, the user may define his own function by using the DEF FN (short for define function) statement. For example,

    10 DEF FNM(X)=X*39.3

defines a function M(X) which converts meters X into inches M(X). The letter following FN is the name of the function and may be any of the 26 letters of the alphabet, thus allowing the user to have up to 26 different functions in a program; i.e., FNA(X), FNB(X), etc.

At this point the user will probably ask what benefit the DEF FN statement has over the LET statement. The answer is that like library functions, once a user defined function has been specified it can be used over and over again without ever having to rewrite it. For example,

    10 DEF FNM(X)=X*39.3
    20 PRINT FNM(10);FNM(10.5);FNM(100);FNM(1000);FNM(10000.6)
    30 END

will cause the computer to print on one line the equivalent in inches of 10, 10.5, 100, 1000, and 10,000.6 meters.

5.6 Strings

In Section 3.7, we saw that variable names can be used to represent numerical quantities. We can just as easily assign variable names to data which are not strictly numerical. This kind of data is often called alphanumeric or alphameric data, and a collection of such data is called a string. In BASIC, a string name must consist of a single alphabetic letter followed by a dollar sign ($); e.g., A$, J$, etc. Furthermore, the computer must be alerted to reserve enough space to store the maximum number of characters including blank spaces, if any, in the string. This is done, as usual, by means of a DIM statement. For example,

    10 DIM A$(15)

tells the computer to reserve space for up to 15 alphameric characters under the string named A. (The user should check the system he is using since string structure varies widely from system to system.) The string length must not exceed 72 characters.

As with numeric variables, we can enter strings into the computer by means of the READ-DATA combination of statements, the LET statement, or the INPUT statement. However, in the case of the DATA and LET statements, the string itself must be enclosed in quotation marks. For this reason, a string may contain any character except the quotation marks. The following program illustrates the use of each of the above statements in entering a string:

```
10 DIM A$(15),B$(15),C$(20)
20 PRINT "WHAT IS YOUR FIRST NAME";
30 INPUT A$
40 LET B$="HI THERE"
50 READ C$
55 DATA "NICE TO MEET YOU"
60 PRINT B$;" ";A$;" ";C$
70 END
```

Note that in the PRINT statement in line 60, we have inserted a blank space in quotation marks between each of the strings. The reason for this is that when a semicolon is used in conjunction with a string in a PRINT statement, no space is left after printing the last character of the string. The results of running this program are as follows:

```
RUN
WHAT IS YOUR NAME?JOHN
HI THERE JOHN NICE TO MEET YOU
DONE
```

We now give another example of the use of strings.

Example 5.6

Suppose we are loading a truck, which has a weight limit of 2500 pounds, piece by piece. Draw a flowchart and write a BASIC program which will check the weight of the next item and print a message indicating whether or not the item can be loaded without exceeding the weight limit.

Flowchart

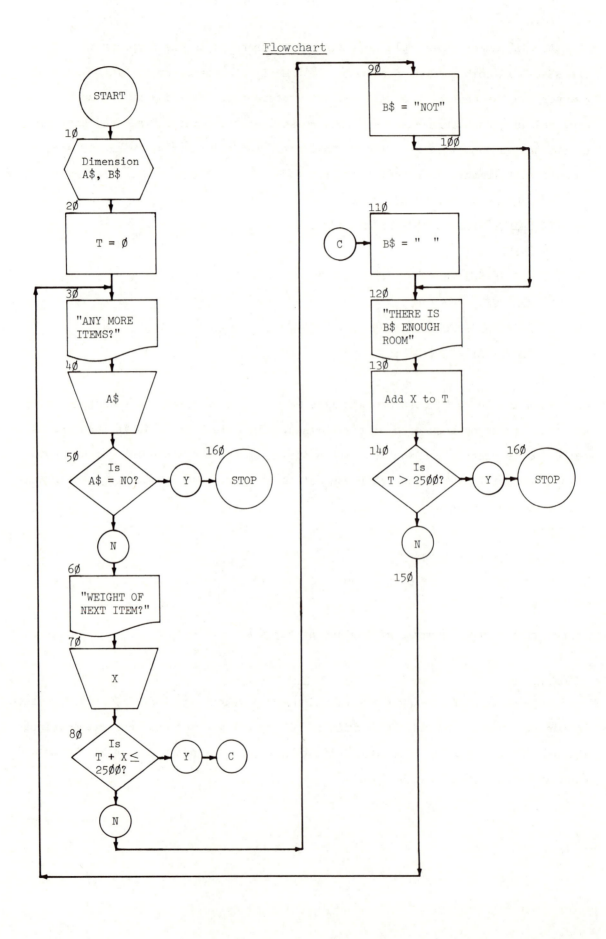

Program

```
1Ø   DIM A$(5),B$(5)

2Ø   LET T=Ø

3Ø   PRINT "ARE THERE ANY MORE ITEMS TO LOAD";

4Ø   INPUT A$

5Ø   IF A$="NO" THEN 16Ø

6Ø   PRINT "WHAT IS THE WEIGHT OF THE NEXT ITEM";

7Ø   INPUT X

8Ø   IF T+X<=25ØØ THEN 11Ø

9Ø   LET B$="NOT"

1ØØ  GOTO 12Ø

11Ø  LET B$=" "

12Ø  PRINT "THERE IS"," ";B$;" ";"ENOUGH ROOM"

13Ø  LET T=T+X

14Ø  IF T>25ØØ THEN 16Ø

15Ø  GOTO 3Ø

16Ø  END
```

Note that, as before, we have presented the flowchart and the program on separate pages because of their length. Note also that in the flowchart, we have introduced a new symbol between the symbol for line 8Ø and the symbol for line 11Ø. This symbol, called a connector, is a small circle

with the same character within it at each end of the connection. Note finally that the single PRINT statement in line 12Ø of the program will type the appropriate message indicating whether or not the item can be loaded without exceeding the weight limit of 25ØØ pounds. This is controlled by the value assigned to the string variable B$ in either line 9Ø or line 11Ø. The results of running this program once are as follows:

```
RUN
ARE THERE ANY MORE ITEMS TO LOAD?YES
WHAT IS THE WEIGHT OF THE NEXT ITEM?2ØØ
THERE IS    ENOUGH ROOM
```

```
ARE THERE ANY MORE ITEMS TO LOAD?YES

WHAT IS THE WEIGHT OF THE NEXT ITEM?5ØØ

THERE IS    ENOUGH ROOM

ARE THERE ANY MORE ITEMS TO LOAD?YES

WHAT IS THE WEIGHT OF THE NEXT ITEM?65Ø

THERE IS    ENOUGH ROOM

ARE THERE ANY MORE ITEMS TO LOAD?YES

WHAT IS THE WEIGHT OF THE NEXT ITEM?45Ø

THERE IS    ENOUGH ROOM

ARE THERE ANY MORE ITEMS TO LOAD?YES

WHAT IS THE WEIGHT OF THE NEXT ITEM?75Ø

THERE IS NOT ENOUGH ROOM

DONE
```

5.7 Subroutines

We saw in Section 5.5 that user defined functions were very helpful in specifying a relation that was to be used many times and at different places in a program. Sometimes we have an entire set of instructions which have to be executed several times in a program at different places. In such cases, it is desirable to organize this set of instructions as a separate part of the program in what is called a "subroutine". Whenever we wish to execute the subroutine we interrupt the natural sequence of the main program and transfer control to the subroutine by means of the GOSUB statement. When the subroutine has been executed, we return control to the main program in the line following the GOSUB by means of the RETURN statement. For example,

    6Ø GOSUB 2ØØ

tells the computer to transfer control to the subroutine beginning in line 2ØØ. In addition, it tells the computer to remember the line number 6Ø as the place in the main program from which we entered the subroutine. Now, the last line in the subroutine would, for example, be

    25Ø RETURN

which would tell the computer that it has completed the execution of the subroutine and should return to the main program at the line immediately following line 6Ø.

The user must be careful to use variable names in the subroutine which are consistent with those used in the main program. In particular, a variable name must have the same meaning in both the main program and the subroutine.

Example 5.7

Draw a flowchart and write a BASIC program which will compute the number of combinations of n objects taken r at a time. Recall that the mathematical formula is

$$C_r^n = \binom{n}{r} = \frac{n!}{r!(n-r)!}$$

where of course $n! = n \cdot (n-1) \cdot (n-2) \cdots 3 \cdot 2 \cdot 1$ for $n \geq 1$ and $0! = 1$. (Thus $1!=1$, $2!=2$, $3!=6$, $4!=24$, etc.) See flowchart and program on following pages.

Note that once again we have presented the flowchart and the program on separate pages because of their length. Note also that in the program the subroutine is accessed at three different places in lines 60, 110, and 140. In line 60, the subroutine is accessed to compute N! for the numerator; in line 110, to compute R! for the denominator; and in line 140, to compute (N-R)! also for the denominator. Note further that there is a STOP statement in line 180. The purpose of this statement is to prevent accidental entry into the subroutine. In addition, this statement is used to terminate execution of the main program; i.e., the program never reaches the END statement in line 260. The results of running this program with five different sets of values for N and R are as follows:

```
RUN
WHAT IS N
?10
WHAT IS R
?7
THE NUMBER OF COMBINATIONS IS 120

DONE
RUN
WHAT IS N
?5
WHAT IS R
?3
THE NUMBER OF COMBINATIONS IS 10
DONE
```

Flowchart

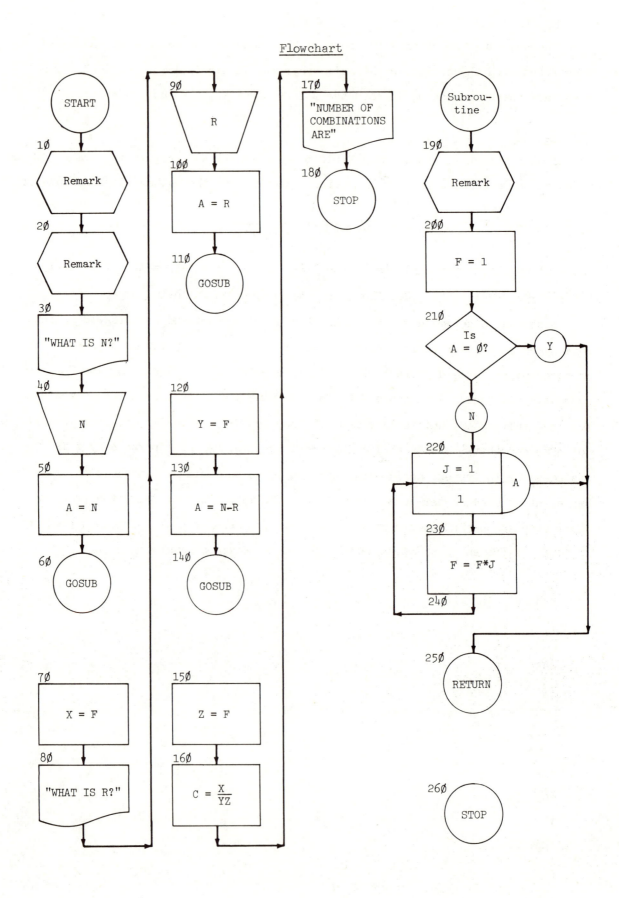

Program

```
10   REM THIS PROGRAM FINDS THE NUMBER OF COMBINATIONS
20   REM OF N THINGS TAKEN R AT A TIME
30   PRINT "WHAT IS N"
40   INPUT N
50   LET A=N
60   GOSUB 200
70   LET X=F
80   PRINT "WHAT IS R"
90   INPUT R
100  LET A=R
110  GOSUB 200
120  LET Y=F
130  LET A=N-R
140  GOSUB 200
150  LET Z=F
160  LET C=X/(Y*Z)
170  PRINT "THE NUMBER OF COMBINATIONS IS",C
180  STOP
190  REM THE SUBROUTINE IS IN LINES 200-250
200  LET F=1
210  IF A=0 THEN 250
220  FOR J=1 TO A
230  LET F=F*J
240  NEXT J
250  RETURN
260  END
```

```
RUN

WHAT IS N
?5
WHAT IS R
?2
THE NUMBER OF COMBINATIONS IS 1Ø


DONE
RUN
WHAT IS N
?4
WHAT IS R
?4
THE NUMBER OF COMBINATIONS IS 1


DONE
RUN
WHAT IS N
?4
WHAT IS R
?Ø
THE NUMBER OF COMBINATIONS IS 1

DONE
```

The user should be aware that for large values of N (typically bigger than 3Ø), the computation of N! causes overflow; i.e., the computed value exceeds the capacity of the machine. Finally, it is usually best to put the subroutine as the last part of the program to prevent accidental entry.

Exercises

Draw a flowchart and write a BASIC program for each of the following.

1. Read in the following B table (a) using only the LET statement, and then
    (b) using the FOR-NEXT loops. Set all entries in the fourth column to zero.

B Table

5	17	8	
84	11	2	
10	31	14	

2.   Read an array A consisting of 100 numbers from a list.   Find and print each of
the following sums:

(a)   A(1) + A(6) + A(11) + ··· + A(96)

(b)   A(2) + A(7) + A(12) + ··· + A(97)

(c)   A(3) + A(8) + A(13) + ··· + A(98)

3.   Read in the following table, then compute the row totals and column totals,
and print out the 5 X 5 table with last column consisting of row totals and
the last row consisting of column totals.

7	3	18	9
17	4	3	88
19	44	31	1
8	5	7	24

4.   Read in the following table, calculate the ratio of each element in the third
column to the item on the same row but in the second column, and print the
ratios along with their row number.

1	14	19	20
8	2	15	18
9	7	3	16
13	10	6	17
12	11	5	4

5.   Print out the table of interest on $1 from 1 to 25 years, at 4%, 5%, 6%, 7%,
and 8% rate of interest.   Your output should have the following format:

```
                         INTEREST AT
          YEAR            4%      5%      6%      7%      8%
            1
            2
            3
            .
            .
```

(Hint: Recall that the interest is $(1+R)^N-1$.)

6. Use an INPUT statement and a function to calculate and print out the volume of a sphere given the radius, where $V=(4/3)\pi R^3$.

7. Print the sine, cosine, and tangent for angles of from 0.01 radians to 1.0 radians in increments of 0.01. On each line print the size of the angle in radians and degrees as well.

8. The RND function can be used for making random assignments to trials like heads vs. tails, up vs. down, etc. Remember that the RND function randomly draws numbers from 0.0 to 0.999999. That is, they are evenly spread over the interval. One could consider therefore all random numbers less than 0.5 to be the equivalent of a head, and those 0.5 or over a tail. Write a program to use the RND function to simulate the tossing of 20 coins, printing the result of each toss as either HEAD or TAIL.

9. Modify the above program to simulate the tossing of 20 coins and then print out the number of HEADS and TAILS.

10. Using the RND function, simulate the coin-tossing game where each player gets $10 to play with. If A and B have the same face showing then A gets a dollar from B. If the faces don't match then A pays. Run the program until one player goes broke, and print the name of the winner.

11. The RND function can be used to define other events such as the face of a die. One way to define the face is to assign a 1 if $RND(X) < 1/6$; a 2 if $1/6 \le RND(X) < 2/6$; a 3 if $2/6 \le RND(X) < 3/6$, etc. (Hint: See Example 5.5 of Section 5.4). Simulate 100 rolls of a die, printing out the face as you go.

12. Modify the above program, so that the faces are tallied and the tally for each face is printed after the 100 tosses.

13. The game of craps is played according to the following rules:

(a) Roll two dice and total the faces. This total is called the point.

(b) If the point is either a 7 or an 11, then you win.

(c) If the point is a 2, 3, or 12, then you lose.

(d) If the point is not the numbers above, but a 4, 5, 6, 8, 9, or 10, then you must reroll.

(e) If the new total is a 7, you lose.

(f) If the new total is your original point, then you win.

(g) If your new total is not your point or a 7, reroll and repeat (e) and (f).

Simulate the game of craps. Then write the BASIC program so that the result of each win and loss is printed, printing the result of each roll.

14. Rewrite the preceding program to simulate the play of 100 games omitting the result of each roll. Then print out the number of wins and losses.

6
Interconversion of Units

To specify a quantitative measurement of any physical quantity, one must do two things:

1 Define the physical quantity being measured.

2 Define the units of measurement.

Thus, in describing the speed of a moving car, one must first have defined speed as the distance traveled per unit time. Then one must state the units of measurement, say, miles per hour. Having made these two specifications, one can now say, "The car is going 55 miles per hour," and be universally understood.

6.1 Simple Conversion Factors

Because the specification of units is purely a matter of definition, it is usually the case that a given physical quantity can be measured in any of several different units. Distance, for example, can be measured in inches, feet, meters, rods, miles, light-years,* etc. In many cases, the scales of the various units are simply proportional to one another, and the zero points on the various scales coincide. For example,

*NOTE: A light-year is the distance light travels in one year and is equal to 5.8781×10^{12} miles.

0 in. = 0 cm = 0 miles

In these cases, the various units are simply related by a multiplicative constant
or conversion factor. It is easy to write a computer program to perform such
conversions.

Example 6.1

Program to convert inches to centimeters (necessary information: 1 in. = 2.54 cm).

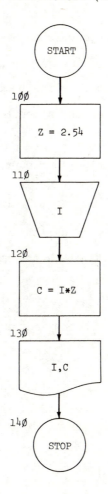

Figure 6.1 Conversion of inches to centimeters.

```
9∅   REM PROGRAM TO CONVERT INCHES TO CENTIMETERS
1∅∅  LET Z=2.54
1∅5  PRINT "ENTER LENGTH IN INCHES";
11∅  INPUT I
12∅  LET C=I*Z
125  PRINT
```

```
13Ø PRINT I;"INCHES= ";C;"CENTIMETERS"
14Ø END
RUN
ENTER LENGTH IN INCHES?5.632
 5.632      INCHES= 14.3Ø53      CENTIMETERS
DONE
```

Here, line 100 establishes Z as the conversion factor. The measurement in
inches is INPUT in line 110, and the conversion of units is performed in line 120.
Note that the nonsignificant digits in the answer have been discarded and that the
precision of the value for Z we build into the program (3 significant figures in
this example) places an upper limit on the number of significant figures in the
answer.

This same program can be used to perform many different conversions just by
changing the value of Z in line 100.

With a little additional effort, we can change the program in Example 6.1 to
convert inches to centimeters or centimeters to inches.

Example 6.2

Program to interconvert inches and centimeters (necessary information:
1 in. = 2.54 cm). (See Figure 6.2.)

```
9Ø  REM PROGRAM TO INTERCONVERT INCHES AND CENTIMETERS
99  DIM L$[2]
1ØØ LET Z=2.54
1Ø8 PRINT "ENTER YOUR MEASUREMENT";
11Ø INPUT X
116 PRINT "CM OR IN";
118 INPUT L$
12Ø IF L$="IN" THEN 16Ø
13Ø LET I=X*(1/Z)
135 PRINT
14Ø PRINT X;"CENTIMETERS= ";I;"INCHES"
15Ø GOTO 18Ø
```

```
16Ø LET C=X*Z

165 PRINT

17Ø PRINT X;"INCHES= ";C;"CENTIMETERS"

18Ø END

RUN

ENTER YOUR MEASUREMENT?1ØØ

CM OR IN?CM

 1ØØ  CENTIMETERS=  39.37Ø1    INCHES

DONE
```

The program now has two possible paths, depending on whether our starting datum is in inches or centimeters. In one case, the conversion factor is just the inverse of the factor in the other case--that is, $1/Z$. If we write the conversion in the form

$$1 \text{ in.} = 2.54 \text{ cm}$$

with the "1" on the left, then for converting inches $\longrightarrow$ centimeters (left to right), $Z = 2.54$. For the reverse conversion, inches $\longleftarrow$ centimeters, $Z = 1/2.54$.

The computer is told which path through the program to take by our INPUTting the units of the starting datum as the string variable I\$ (in line 118) when the program is RUN.

When writing a program that may be used by other persons, it is desirable to have the computer print labels indicating the nature of any input required; this is the purpose of the PRINT statements in lines 108 and 116. These labeling statements are not included in the flowchart.

Problem 6.1

Modify the program in Example 6.1 to convert pounds into grams (1 lb = 453.6 g).

Problem 6.2

Write a program to convert liquid volumes in ounces to liters (1 oz = 1/32 qt, 1 qt = 0.9463 liter).

Problem 6.3

Write a program to prepare a table of lengths in inches and the corresponding lengths in centimeters. Have the table run from 0 to 40 in. in intervals of 1 in.

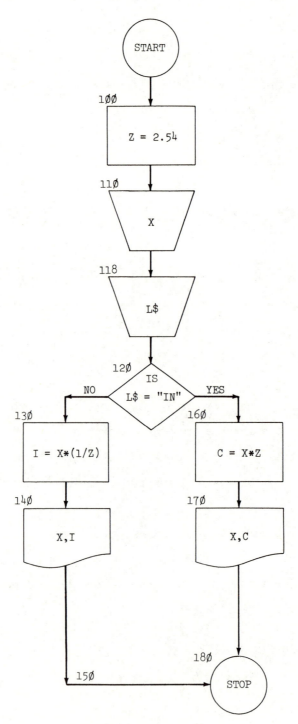

Figure 6.2   Interconversion of inches and centimeters.

In some cases, a conversion of units may involve more than one conversion step. Suppose we wish to convert years into seconds. Because the computer is doing the arithmetical chores, it is easy for us to set up a program involving several conversions based on the following familiar information:

$$1 \text{ yr} = 365 \text{ days}$$

$$1 \text{ day} = 24 \text{ hr}$$

$$1 \text{ hr} = 60 \text{ min}$$

$$1 \text{ min} = 60 \text{ sec}$$

Because the conversion will be years→days→hours→minutes→seconds (left to right in each of the above equivalences), the conversion factors will just be the numbers on the right-hand side of the equivalences. This leads to a straightforward program.

Example 6.3

Program to convert years to seconds (necessary information: 1 yr = 365 days, 1 day = 24.0 hr, 1 hr = 60.0 min, 1 min = 60.0 sec). (See Figure 6.3.)

```
9Ø  REM PROGRAM TO CONVERT YEARS TO SECONDS
1ØØ LET Z1=365
11Ø LET Z2=24
12Ø LET Z3=6Ø
13Ø LET Z4=6Ø
138 PRINT "ENTER DATUM IN YEARS";
14Ø INPUT Y
15Ø LET S=Y*Z1*Z2*Z3*Z4
16Ø PRINT Y;"YEARS= ";S;"SECONDS"
17Ø END
RUN
ENTER DATUM IN YEARS?1
 1     YEARS=  3.1536ØE+Ø7    SECONDS
DONE
RUN
ENTER DATUM IN YEARS?5.5
 5.5        YEARS=  1.73448E+Ø8    SECONDS
DONE
```

Writing the program in Example 6.3 would be of questionable value if one needed to convert only one piece of data. It would be quicker to do the calculation

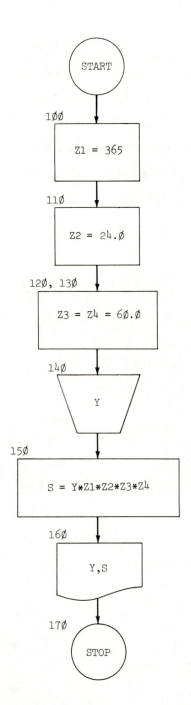

Figure 6.3 Conversion of years to seconds.

on a desk calculator. However, the job soon becomes an appropriate computer task if you are asked, for some obscure reason, to prepare a table of years-to-seconds equivalences for, say, 0 through 10 years. A tedious undertaking on a calculator, this job is easily done by the computer if we add two lines to the program in

Example 6.3 and remove two others. That is, remove lines 138 and 140 and add

140 FOR Y=1 TO 10

165 NEXT Y

In practice, we would also change the output format to a form more suitable for a table. These modifications are shown in Example 6.4.

Example 6.4

Program to prepare years-to-seconds table (necessary information: same as in Example 6.3). (See Figure 6.4.)

```
9∅  REM PROGRAM FOR YEARS-TO-SECONDS TABLE
1∅∅ LET Z1=365
11∅ LET Z2=24
12∅ LET Z3=6∅
13∅ LET Z4=6∅
135 PRINT "YEARS";" SECONDS"
14∅ FOR Y=1 TO 1∅
15∅ LET S=Y*Z1*Z2*Z3*Z4
16∅ PRINT Y;S
165 NEXT Y
17∅ END
RUN
YEARS SECONDS
   1      3.1536∅E+∅7
   2      6.3∅72∅E+∅7
   3      9.46∅8∅E+∅7
   4      1.26144E+∅8
   5      1.5768∅E+∅8
   6      1.89216E+∅8
   7      2.2∅752E+∅8
   8      2.52288E+∅8
   9      2.83824E+∅8
  1∅      3.1536∅E+∅8
DONE
```

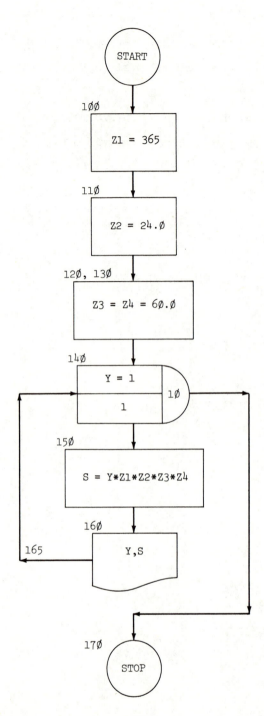

Figure 6.4 Modification of conversion of years to seconds to produce table.

6.2 Temperature Conversion

    The examples we have dealt with so far have all involved simple multiplicative

conversion factors. In the case of temperature-scale conversions, however, the zeros of the various scales do not coincide.

$$0^{\circ}C \neq 0^{\circ}K \neq 0^{\circ}F$$

Thus, the conversion of temperature units is somewhat more complicated than our previous conversions.

Since 32° on the Fahrenheit scale is equivalent to 0° on the centigrade scale (both these scale readings represent the same physical temperature), and since $212^{\circ}F$ is equivalent to $100^{\circ}C$, we can draw two adjacent thermometers as in Figure 6.5.

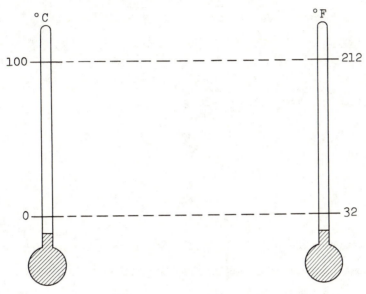

Figure 6.5   Parallel centigrade and Fahrenheit thermometers.

If one were to mark off each of the thermometers in Figure 6.5 in degrees, there would be 100 degrees between the two fixed points on the centigrade thermometer and 180 degrees between the corresponding physical points on the Fahrenheit thermometer. In any physical temperature interval, the number of Fahrenheit degrees will always be 1.8 times the number of centigrade degrees. Therefore, if we now consider a physical temperature such that the centigrade thermometer reads $1^{\circ}C$, we will have the situation depicted in Figure 6.6.

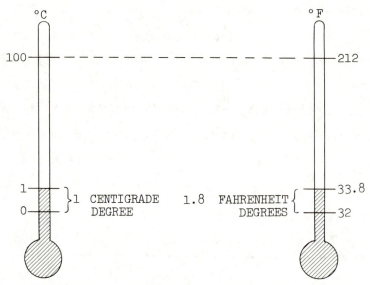

Figure 6.6   Parallel thermometers reading $1°C$ and $33.8°F$.

The one-degree temperature interval on the centigrade thermometer corresponds to a
1.8-degree interval on the Fahrenheit thermometer, and the actual Fahrenheit temperature is $1.8° + 32° = 33.8°F$.   In general,

$$°F = 1.8 \times °C + 32° \tag{6.1}$$

If you forget this formula, you can always draw the two thermometers in Figure 6.6
and derive the correct relationship as we have done.

Equation (6.1) can be rearranged to give the reverse conversion

$$°C = \frac{1}{1.8} \times (°F - 32°) \tag{6.2}$$

One other temperature scale of importance is the Kelvin or absolute scale.
This is related to the centigrade scale as

$$°K = °C + 273.15° \tag{6.3}$$

(It should be noted that the constants appearing in Equations (6.1) to (6.3) are
defined values and thus may be considered to have an unlimited number of significant
figures.)

The programming of temperature conversions does not differ from the other
examples we have done, except that the conversion steps now involve addition as well

as multiplication.   Example 6.5 shows such a program.

Example 6.5

   Program to interconvert Kelvin and Fahrenheit temperatures.

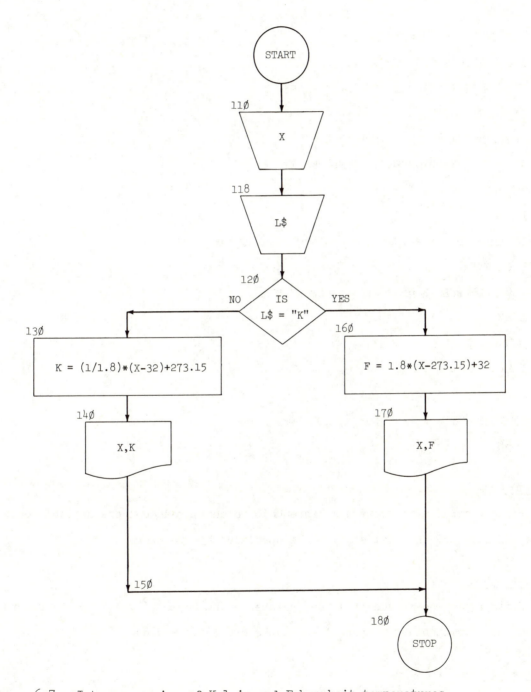

Figure 6.7   Interconversion of Kelvin and Fahrenheit temperatures.

```
1ØØ REM PROGRAM FOR K/F TEMPERATURE CONVERSION
1Ø8 PRINT "ENTER TEMPERATURE";
11Ø INPUT X
116 PRINT "K OR F";
118 INPUT L$
12Ø IF L$="K" THEN 16Ø
13Ø LET K=(1/1.8)*(X-32)+273.15
14Ø PRINT X;"DEGREES F =";K;"DEGREES K"
15Ø GOTO 18Ø
16Ø LET F=1.8*(X-273.15)+32
17Ø PRINT X;"DEGREES K =";F;"DEGREES F"
18Ø END
RUN
ENTER TEMPERATURE?32
K OR F?F
 32   DEGREES F = 273.15     DEGREES K
DONE
RUN
ENTER TEMPERATURE?Ø
K OR F?K
 Ø    DEGREES K =-459.67     DEGREES F
DONE
```

Problem 6.4

Write a program to prepare a Fahrenheit to centigrade conversion table over any range and with any interval size, to be specified by the user.

Problem 6.5

Write a program to accept a temperature in any scale ($^{\circ}$F, $^{\circ}$C, or $^{\circ}$K) and to output the corresponding temperatures in the other two scales.

Problem 6.6

(a) Write a program that will compare the Fahrenheit and corresponding centigrade temperature over the range -50°F to +50°F and that will find, to the nearest

degree F, the Fahrenheit temperature at which the two temperature scales readings
are numerically closest together. You will probably want to use the ABS function.

(b) Solve this same problem algebraically, by hand. Note the essential difference in the computer approach versus the algebraic method.

6.3 Angle Conversion

The interconversion of angle measurements between degrees and radians is not
commonly encountered in manual chemical computations. Whenever we deal with angles,
we work in the familiar units of degrees. Most computers, however, deal with angles
only in radians. For example, the predefined BASIC function

$$SIN(X)$$

assumes that X is in radians, not degrees. Thus, the chemist-programmer must often
include in his programs small routines that convert angles in degrees (for the
user's convenience) to radians (for the computer's use) and back.

This conversion is easily programmed, using our previous techniques, if one
makes use of the following information:

$$360^{\circ} = 1 \text{ circle} = 2\pi \text{ radians}$$

or

$$1^{\circ} = 2\pi/360 \text{ radians}$$

In converting from degrees to radians, for example, one traverses this equivalence
from left to right; therefore the conversion factor is $2\pi/360$. The value of π
must be explicitly supplied by the programmer, and because we do not want this value
to limit the precision of our computation, we must supply it with the maximum number
of significant figures the computer will allow. To six places,

$$\pi = 3.14159$$

6.4 Logarithm Conversion

A similar situation exists when dealing with computations involving logarithms.
Many scientific calculations use logarithms to the base e (natural logarithms), but
some important expressions are most commonly formulated with base 10 logarithms
(common logarithms). Computers usually deal exclusively with base e logarithms.

Formulation of the general conversion procedure for logarithms is an interesting algebraic exercise (done better by a person than by a computer). Suppose you
want the log of a number, N, to two different bases, say e and 10. Then

$$x = \log_e N$$

$$y = \log_{10} N$$

By the definition of logarithms,

$$N = e^x$$

$$N = 10^y$$

so

$$e^x = 10^y$$

If we take the log to the base e of both sides of this equation, we have

$$x \log_e e = y \log_e 10$$

but

$$\log_e e = 1$$

therefore

$$x = y \log_e 10$$

Thus, to convert from the base 10 log of N to the base e log, we multiply the former by $\log_e 10$, the log of the old base to the new base we are interested in. This conversion factor is called the "modulus" of the two bases.

Problem 6.7

Crystals can be considered to be composed of atoms or molecules arranged in orderly sets of planes. The spacing between these planes can be measured indirectly using x-ray diffraction. An x-ray beam is directed at the crystal and is diffracted by the crystal at some angle relative to the original beam, and this angle is known as 2θ, which can be measured for various orientations of the crystal. The spacings d between the planes causing a particular diffracted beam can be found from Bragg's equation

$$d = \frac{\lambda}{2 \sin\theta}$$

Here, d is the plane spacing in $\overset{o}{A}$, λ is the wavelength of the x-rays in $\overset{o}{A}$, and θ is $\frac{1}{2}(2\theta)$.

Write a program to compute the interplanar spacing of any set of crystal planes, given the corresponding 2θ. Assume $\lambda = 1.5418 \overset{o}{A}$. Use your program to compute the spacings for the following sets of planes in a diamond crystal. The sets are referred to by their Miller indices (h,k,l).

Set of planes (hkl)	2θ, degrees ($\lambda = 1.5418\text{Å}$)
1 1 1	43.969
2 0 0	51.223
2 2 0	75.369
3 1 1	91.587
2 2 2	96.957
4 0 0	119.66
3 3 1	140.82
4 2 0	150.29

Problem 6.8

Write a program to prepare a table of natural and common logarithms of numbers from 1 to 10 in increments of 0.1.

7

Stoichiometry

Suppose someone hands you a small jar full of red powder and says, "I don't know what this is, but it's 90.7 percent lead and 9.33 percent oxygen."* Finding out the chemical formula for a substance is a typical problem in stoichiometry. Stoichiometry is the area of chemistry that translates the macroscopic, laboratory-scale measurement of masses into information about the (sub)microscopic world of atoms.

Stoichiometry calculations are based on the mole concept--the idea that 1 gram-atomic-weight of any element contains 6.024×10^{23} atoms, or that 1 gram-molecular-weight of any compound contains 6.024×10^{23} molecules. Nearly all stoichiometric computations are variations on this theme. Balancing chemical equations is also part of stoichiometry, but this type of calculation is more efficiently done by hand than by computer.

7.1 Moles and Percent Composition

The simplest kind of stoichiometric problem requires finding the number of moles of a substance corresponding to a given mass. This is accomplished by dividing the given mass by the atomic weight, or by the molecular weight, of the substance, as the case may require. Such calculations are easily programmed, as in Example 7.1.

*NOTE: Because of experimental error, the percentages may not always add up to 100 percent.

Example 7.1

Program to calculate the number of moles and atoms in any mass of silver (necessary information: gram-atomic-weight of silver (M1) = 107.870g/mole, Avogadro's number (N) = 6.0238×10^{23} atoms/mole).

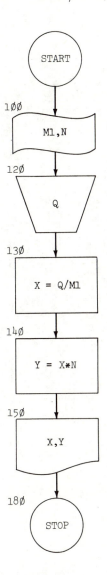

Figure 7.1   Calculation of moles and atoms in any mass of silver.

```
9Ø  REM PROGRAM.TO COMPUTE # OF MOLES AND ATOMS
1ØØ READ M1,N
11Ø PRINT "ENTER MASS OF SILVER (GRAMS)";
12Ø INPUT Q
13Ø LET X=Q/M1
```

```
14Ø LET Y=X*N
15Ø PRINT X;"MOLES",Y;"ATOMS"
16Ø DATA 1Ø7.87
17Ø DATA 6.Ø24E+23
18Ø END
RUN
ENTER MASS OF SILVER (GRAMS)?1Ø7.87
 1     MOLES      6.Ø24ØØE+23    ATOMS
DONE
RUN
ENTER MASS OF SILVER (GRAMS)?215.74
 2     MOLES      1.2Ø48ØE+24    ATOMS
DONE
RUN
ENTER MASS OF SILVER (GRAMS)?1Ø
 9.27Ø42E-Ø2   MOLES           5.5845ØE+22    ATOMS
DONE
```

Here, the necessary information is READ in line 100. The measured mass (in grams) is INPUT from the keyboard in line 120. The number of moles and number of atoms are calculated in lines 130 and 140, and the results are printed in line 150. By changing line 130 from a division to a multiplication, deleting line 140, and appropriately changing the PRINT statements, this program can be made to find the mass of a given number of moles of silver; this is, of course, just the reverse of the problem we have treated.

The program can be modified to work for any element (or compound) by changing the value in the DATA statement 160 to the appropriate atomic weight (or molecular weight).

Problem 7.1
    Modify the program in Example 7.1 to find the following information:
    (a) The number of atoms in 12 g nitrogen
    (b) The number of atoms in 35.4 g chlorine
    (c) The number of molecules in 30 g ethane (C_2H_6)

Problem 7.2

Change the program in Example 7.1 to calculate the number of grams corresponding to a given number of moles of an element. Compute the mass of

(a) 3.25 moles of helium

(b) 3.25 moles of argon

(c) 3.25 moles of gold

A particularly straightforward problem is the calculation of the percent composition for a given material. If a 10-g sample of a compound contains 3 g of
element X and 7 g of element Y, the percent composition is clearly 30 percent X and
70 percent Y. Conversely, if one knew the composition to be 30 percent X and 70
percent Y, one would know that, say, a 10-g sample would contain 3 g of element X

Example 7.2

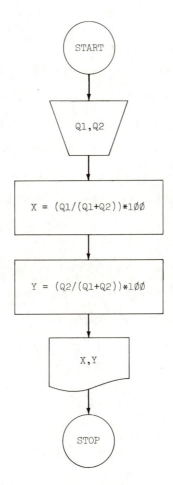

Figure 7.2   Flowchart for computing percent composition of binary compounds given
              the mass analysis.

and 7 g of element Y. A flowchart for computing such percentage compositions for binary compounds is shown in Example 7.2 on the preceding page.

Here, Q1 and Q2 represent the masses of the two elements, and X and Y are the respective percentages.

Problem 7.3

Convert the flowchart in Example 7.2 into a program. Calculate the percent composition of each of the following substances:

(a) 13.88 g lithium;   16.00 g oxygen

(b) 0.1468 g nickel;   0.3995 g bromine

(c) 1.401 g cerium;   0.3200 g oxygen

(d) 1.401 g cerium;   0.2400 g oxygen

(e) 0.1401 g nitrogen; 0.03024 g hydrogen

7.2   Chemical Formulas--Ionic Compounds

One can also calculate the percentage composition of a material from its chemical formula. For example, one mole of PbO_2 contains one mole of Pb and two moles of O. The respective masses of these two components could be calculated with the suggested reverse version of the program in Example 7.1. The resulting masses could be INPUT into a program based on Example 7.2, which would compute the percent composition. These two tasks are both performed in the following composite program.

Example 7.3

Program to calculate percent composition of lead-oxygen compounds, given the chemical formula (necessary information: gram-atomic-weight of lead (M1) = 207.19 g/mole, gram-atomic-weight of oxygen (M2) = 15.999 g/mole).

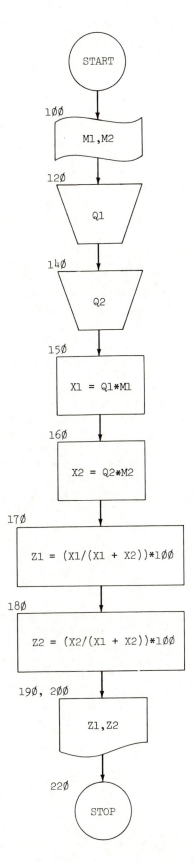

Figure 7.3   Calculation of percent composition from formula.

```
90    REM PROGRAM TO COMPUTE % COMPOSITION
95    REM FROM CHEMICAL FORMULA
100   READ M1,M2
110   PRINT "ENTER LEAD SUBSCRIPT";
120   INPUT Q1
130   PRINT "ENTER OXYGEN SUBSCRIPT";
140   INPUT Q2
150   LET X1=Q1*M1
160   LET X2=Q2*M2
170   LET Z1=(X1/(X1+X2))*100
180   LET Z2=(X2/(X1+X2))*100
190   PRINT Z1;"% LEAD"
200   PRINT Z2;"% OXYGEN"
205   REM ATOMIC WEIGHTS
210   DATA 207.19,15.999
220   END
RUN
ENTER LEAD SUBSCRIPT?2
ENTER OXYGEN SUBSCRIPT?1
 96.2826    % LEAD
 3.71742    % OXYGEN
DONE

RUN
ENTER LEAD SUBSCRIPT?2
ENTER OXYGEN SUBSCRIPT?3
 89.6195    % LEAD
 10.3805    % OXYGEN
DONE
```

Although different variable names have been used, the functioning of this program is just like that of Examples 7.1 and 7.2.

Problem 7.4

Use the program in Example 7.3 to find the percent composition of each of the following compounds:

(a) PbO

(b) Pb_2O_3

(c) PbO_2

(d) Pb_3O_4

(e) Pb_2O

Problem 7.5

Modify the program from Example 7.3 to find the percent composition of each of the following compounds:

(a) NO_2

(b) NO

(c) N_2O

(d) N_2O_5

Problem 7.6

Modify the program from Example 7.3 to find the percent composition of each of the following compounds. Compare the results with your output from Problem 7.3.

(a) Li_2O

(b) $NiBr_2$

(c) CeO_2

(d) Ce_2O_3

(e) NH_3

You will need to change the DATA statements for the different elements involved.

Of much chemical importance is the job that is the reverse of that performed in Example 7.3. That is, we are given as analytical data a percentage composition (such as 69.9 percent iron, 30.1 percent oxygen) and are asked to find the corresponding chemical formula for the material in question. This is a straightforward procedure when done by hand; computerizing it will require some interesting programming tricks. The steps involved in this type of problem are as follows:

1 Assume a 100-g sample, so that the percentages of each element can be regarded as grams. In our example, 69.9 percent Fe thus becomes 69.9 g of iron, and 30.1 percent O becomes 30.1 g of oxygen.

2 Divide the mass in grams of each element by the respective atomic weight of that element.

$$69.9 \text{ g Fe}/(55.847 \text{ g/mole}) = 1.25 \text{ mole Fe}$$
$$30.1 \text{ g O}/(15.999 \text{ g/mole}) = 1.88 \text{ mole O}$$

This step gives numbers that could be used as subscripts in a chemical formula, except that they are not the "small whole numbers" required by convention.

3 We therefore divide each number from Step 2 by the smallest of those numbers. In some cases, this division will give the sought-for integers.

$$1.25/1.25 = 1$$
$$1.88/1.25 = 1.5$$

If the results are not integers, we proceed to Step 4.

4 Multiply the results from Step 3 by successive integers--2, 3, 4 ...--until an integer is found that, when multiplied times the Step 3 results, gives whole numbers. These whole numbers are the subscripts for the chemical formula.

$$2 \times 1 = 2$$
$$2 \times 1.5 = 3$$

Formula is Fe_2O_3.

If one starts with real analytical data, which is subject to various errors, then exact whole numbers will usually not be obtained by this procedure. Thus, it must be decided how close a number must be to the nearest whole number in order actually to qualify as a whole number. Clearly, it is incorrect to round off $FeO_{1.5}$ to FeO. However, it is usually safe to assume that a result like $Fe_2O_{2.99}$ is the compound Fe_2O_3. We might consider, somewhat arbitrarily, that if a subscript is within 0.1 of the nearest integer, it can be rounded to that integer. For calculations based on very good analytical data, one would want to make this tolerance smaller.*

The program in Example 7.4 solves this type of problem.

Example 7.4

Program to calculate chemical formula from percentage composition for lead-oxygen compounds (necessary information: gram-atomic-weight of lead = 207.19 g/mole, gram-atomic-weight of oxygen = 15.999 g/mole). (See Figure 7.4.)

```
9Ø   REM PROGRAM TO FIND MOLECULAR FORMULA
1ØØ  READ M1,M2
1Ø5  LET J=2
11Ø  PRINT "ENTER % LEAD   ";
12Ø  INPUT P1
13Ø  PRINT "ENTER % OXYGEN";
14Ø  INPUT P2
15Ø  LET A1=P1/M1
16Ø  LET A2=P2/M2
17Ø  IF A1>A2 THEN 2ØØ
18Ø  LET A2=A2/A1
185  LET A1=A1/A1
19Ø  GOTO 21Ø
2ØØ  LET A1=A1/A2
2Ø5  LET A2=A2/A2
21Ø  LET A8=A1
22Ø  LET A9=A2
23Ø  IF ABS(A1-INT(A1+.5))>.1 THEN 28Ø
24Ø  IF ABS(A2-INT(A2+.5))>.1 THEN 28Ø
```

*NOTE: There <u>are</u> compounds with noninteger formulas, like $MnO_{1.95}$.

```
25Ø PRINT "PB";INT(A1+.5)
26Ø PRINT "O ";INT(A2+.5)
27Ø STOP
28Ø LET A1=A8*J
29Ø LET A2=A9*J
3ØØ LET J=J+1
31Ø GOTO 23Ø
4ØØ REM ATOMIC WEIGHTS
4Ø1 DATA 2Ø7.19,15.999
5ØØ END
RUN
ENTER % LEAD   ?9Ø.7
ENTER % OXYGEN? 9.3
PB 3
O  4
DONE
RUN
ENTER % LEAD   ?96.3
ENTER % OXYGEN? 3.7
PB 2
O  1
DONE
```

In effect, this program carries out the same steps that one would perform by hand in solving such a problem. Although Step 1 is not programmed (because it results in no actual change in the numbers we are working with), Step 2 is performed by lines 150 and 160. Lines 170 to 205 find out which quotient is smaller (A1 or A2) and take the appropriate action as in Step 3. Lines 280 through 300 perform Step 4, if necessary.

Of special interest are lines 230 and 240. They decide whether Step 4 (lines 280 to 300) is indeed necessary. This is accomplished with a common programming trick employing the INT function. The expression $INT(Z+.5)$ gives the value of Z rounded off to the nearest integer. Thus,

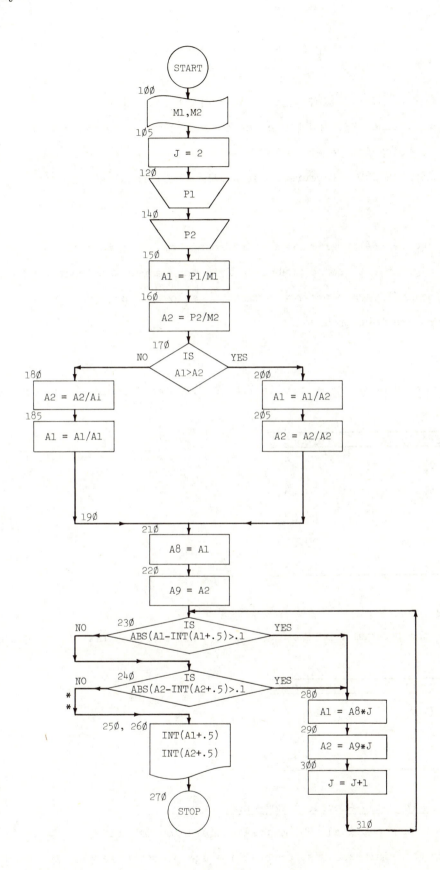

Figure 7.4 Calculation of formula from percent composition.

$$INT(3.45 + .5) = INT(3.95) = 3$$

while

$$INT(3.55 + .5) = INT(4.05) = 4$$

In this manner, lines 230 and 240 check to see if the subscripts A1 and A2 are within 0.1 of being integers.

Lines 250 and 260, of course, output the final results.

Problem 7.7

Write and test a program segment to round numbers off to the nearest tenth. Modify the algorithm to round numbers off to the nearest hundredth. Generalize the algorithm and modify your program segment to allow the user to specify the place to which numbers are to be rounded.

Problem 7.8

Modify the program from Example 7.4 to find the formulas of the following compounds:

	Percent Fe	Percent S
(a)	63.53	36.47
(b)	46.55	53.45
(c)	53.73	46.27
(d)	60.38	39.62
(e)	56.64	43.36

Problem 7.9

Modify the program from Example 7.4 to find the formulas of the following compounds:

	Percent Fe	Percent O
(a)	77.73	22.27
(b)	69.94	30.06
(c)	72.36	27.64

7.3 Chemical Formulas--Covalent Compounds

The procedure outlined in the previous section is fine for any ionic compound, because such materials are just crystalline agglomerates of roughly spherical ions.

There is no chemical difference between NaCl and Na_6Cl_6; by convention, we prefer the simpler formula.

For covalent compounds, where the fundamental pieces of the material are molecules with definite geometric identity, a further consideration arises. For, although NO_2 and N_2O_4 have the same percentage composition (30.45 percent N, 69.55 percent O), they are quite different molecules and have very different properties. This means that, for covalent (molecular) compounds, a given percentage composition does not necessarily lead to the correct molecular formula. Using the procedure in Example 7.4, the data 30.45 percent N, 69.55 percent O would lead to the formula NO_2, whether it was NO_2 or N_2O_4 that had been analyzed. To find the correct formula for covalent compounds, an additional piece of information is needed: the molecular weight. The molecular weight of NO_2 is 46.005, while that of N_2O_4 is 92.010. This complication leads to one additional step that is carried out for determining formulas of covalent compounds:

5   Divide the given molecular weight by the weight of the formula unit resulting from Step 3 or Step 4 and multiply the subscripts by this quotient.

This step can be inserted into our existing program. The following flowchart would be inserted between the asterisks in the flowchart of Example 7.4. Of course, the Pb-O case is no longer a suitable example because lead-oxygen compounds are ionic.

Example 7.5

Modifications to Example 7.4 needed to compute formulas of covalent compounds [additional necessary information: molecular weight of compound (W)]. (See Figure 7.5.)

```
242 PRINT "ENTER M.W.";
243 INPUT W
244 LET Z=A1*M1+A2*M2
245 LET Q=W/Z
246 LET A1=A1*Q
247 LET A2=A2*Q
```

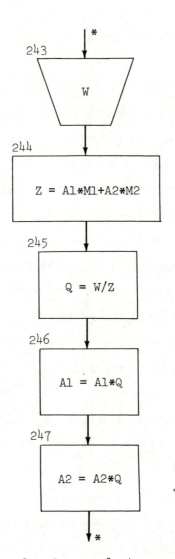

Figure 7.5 Computation of formulas for covalent compounds.

Problem 7.10

Use the program suggested in Example 7.5 to find the molecular formulas of the following covalent carbon-hydrogen compounds:

	Percent C	Percent H	Molecular weight
(a)	92.26	7.74	78.11
(b)	92.26	7.74	26.04
(c)	85.64	14.36	56.10
(d)	85.64	14.36	28.05
(e)	83.63	16.37	86.17
(f)	93.71	6.29	128.16

Remember to change the atomic weight data in the program to the appropriate values for carbon and hydrogen.

7.4 Information Retrieval and Data Bases

The programs in the previous examples in this chapter are useful and, once written, can save a significant amount of manual computational labor. However, the utility of these programs is limited by the fact that the programs as presented can deal only with a small group of compounds--say, the lead-oxygen compounds as in Example 7.4. Would it not be advantageous to construct the programs so that they would have the information necessary for working with, for example, any binary compound?

Information retrieval is a computer application in which a large amount of data is available to the computer and can be accessed and manipulated according to the needs of the user. If we extend our stoichiometric programs to include information retrieval features, we will greatly enhance their usefulness. Specifically, we would like to be able simply to indicate to the computer at RUN time which two chemical elements are involved in our problem and not have to modify the program itself for each new problem.

The store of information in an information retrieval system is called the "data base." In our present application, the data base would include the symbols for all the chemical elements and the corresponding atomic weights. In the problem posed in Example 7.4, for instance, we would like to INPUT the symbols of the elements involved and their respective percentages and have the computer choose from its data base the appropriate atomic weights for the calculation.

There are various ways to approach information retrieval problems. The way in which the data base is to be organized is a question of major importance. Two primary kinds of data base are the "random access" and the "sequential access" types. The difference between them can be appreciated by considering an example. Suppose that you have stored a data base consisting of a list of 100 pieces of data and that you need to retrieve the nineteenth item in the list. If the data base is organized as a random access file, the computer can go directly to the nineteenth item in the list and retrieve it. With a sequential access file, the computer would have to begin at item one in the list and actually access each successive item, keeping count

of the number of items accessed, until it came to the nineteenth item.   There is an
important trade-off in the choice between random and sequential access to a data
base.   Random access is clearly faster, especially for large data bases; but sequential access can frequently allow much more compact storage of the data.

Let us extend the program in Example 7.4 by creating a data base that includes
the first twenty elements in the Periodic Table.   While large data bases are usually
stored in computer peripherals, such as magnetic disks, we will store our small one
in a series of DATA statements.   We shall employ a type of sequential access to this
data base.

Example 7.6

Extension of formula calculating program to the first twenty elements.   (See
Figure 7.6.)

```
5    DIM A$[2],Y$[2],Z$[2]
10   PRINT "ENTER SYMBOL FOR ELEMENT 1";
15   INPUT Y$
20   PRINT "ENTER SYMBOL FOR ELEMENT 2";
25   INPUT Z$
30   LET I=0
35   IF I=2 THEN 105
40   READ A$,M
45   IF A$=Y$ THEN 60
50   IF A$=Z$ THEN 75
55   GOTO 40
60   LET M1=M
65   LET I=I+1
70   GOTO 35
75   LET M2=M
80   LET I=I+1
85   GOTO 35
100  REM   REMOVE LINE 100 FROM EXAMPLE 7.4
110  PRINT "ENTER % ";Y$;
130  PRINT "ENTER % ";Z$;
250  PRINT Y$;INT(A1+.5)
```

```
26Ø PRINT Z$;INT(A2+.5)
4Ø1 DATA "H",1.ØØ8
4Ø2 DATA "HE",4.ØØ26
4Ø3 DATA "LI",6.939
4Ø4 DATA "BE",9.Ø122
4Ø5 DATA "B",1Ø.811
4Ø6 DATA "C",12.Ø11
4Ø7 DATA "N",14.Ø77
4Ø8 DATA "O",15.999
4Ø9 DATA "F",18.998
41Ø DATA "NE",2Ø.183
411 DATA "NA",22.99
412 DATA "MG",24.312
413 DATA "AL",26.982
414 DATA "SI",28.Ø86
415 DATA "P",3Ø.974
416 DATA "S",32.Ø64
417 DATA "CL",35.453
418 DATA "AR",39.948
419 DATA "K",39.1Ø2
42Ø DATA "CA",4Ø.Ø8
```

The modifications consist of (a) addition of the information retrieval section; (b) changes to the labeling statements; (c) addition of the data base itself. The latter two changes are self-explanatory.

To achieve retrieval of the proper atomic weights, the symbols of the desired elements are INPUT as string variables in lines 15 and 25. The successive items in the data base are then READ in as (symbol, atomic weight) pairs in line 40. Each symbol is tested to see if that element is one of those requested for the problem (lines 45 and 50). If not, the next element is READ from the data base.

If the element is needed, then M1 or M2 (same as in Example 7.4) is set equal to the atomic weight for use in the subsequent calculation. The index I is used to indicate when both of the needed elements have been found so that we don't waste time going through the whole data base unnecessarily.

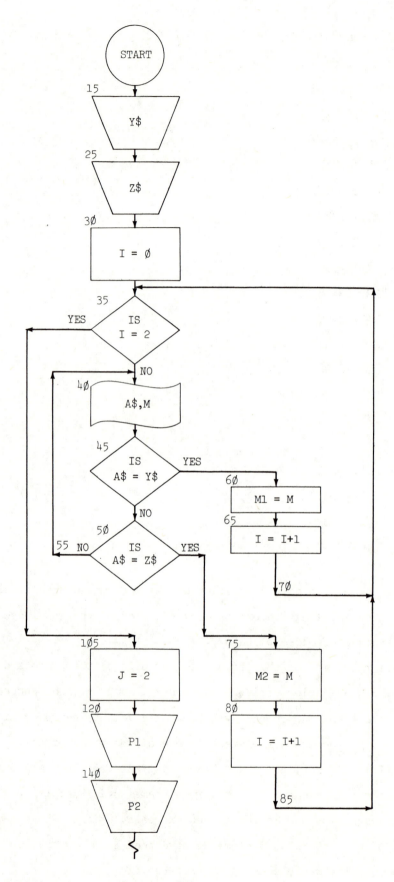

Figure 7.6 Modifications to Example 7.4.

Problem 7.11

Use the modified program suggested in Example 7.6 to find the formulas of the following ionic compounds:

	Weight percent of cation		Weight percent of anion	
(a)	Mg	25.53	Cl	74.47
(b)	Ca	36.11	Cl	63.89
(c)	Na	39.33	Cl	60.67
(d)	K	52.44	Cl	47.56
(e)	Ca	62.53	C	37.47
(f)	Na	58.91	S	41.09
(g)	Na	74.19	O	25.81
(h)	Li	26.75	F	73.25

Problem 7.12

Extend the data base in Example 7.6 to include Fe, Cu, Pb, and Hg. Use the program to find the formulas of the following compounds:

	Weight percent of cation		Weight percent of anion	
(a)	Hg	73.88	Cl	26.12
(b)	Hg	92.60	O	7.40
(c)	Hg	96.16	O	3.84
(d)	Cu	66.46	S	33.54
(e)	Cu	79.85	S	20.15
(f)	Cu	75.47	P	24.53
(g)	Cu	93.15	N	6.85
(h)	Cu	94.08	O	5.92
(i)	Cu	88.82	O	11.18
(j)	Cu	79.88	O	20.12
(k)	Fe	78.28	P	21.72
(l)	Fe	34.43	Cl	65.57
(m)	Fe	49.49	F	50.51
(n)	Pb	71.14	N	28.86
(o)	Pb	59.36	Cl	40.64
(p)	Pb	74.50	Cl	25.50

8
Gases

Following birth, we are completely surrounded by gas every minute of our lives, with the exception of those few individuals who go in for skin-diving, lunar missions, and the like. And yet, because of their invisibility, gases seem to most of us somewhat mysterious and elusive. Nonetheless, the properties of gases are more easily understood than those of liquids or of solids.

8.1 Ideal Gases

A great deal of insight into the behavior of gases can be gained by studying an imaginary material known as an ideal gas. The two attributes of an ideal gas that guarantee its nonexistence are as follows:

1 A molecule of an ideal gas occupies no volume (although it has mass).

2 The molecules of an ideal gas neither attract nor repel one another.
These hypothetical qualities cannot be found in real gases because real molecules, of course, do occupy volume, and they are capable of exerting forces on one another. However, it is true that the model of an ideal gas closely approximates the behavior of real gases at relatively low pressures at relatively high temperatures. Under these conditions, the molecules are so far apart and move so fast that their size and attractive forces do not significantly affect the behavior of the gas.

The "state" of a sample of an ideal gas is determined by the number of moles (n) present, the pressure (P), volume (V), and temperature (T). These variables are interrelated by the "ideal gas equation"

$$PV = nRT$$

where R is a constant whose units and numeric value are determined by the units used to measure the other variables. For example:

for P in atmospheres
     V in liters
     n in moles
     T in ^{o}K
} $R = 0.08205$ liter·atm/^{o}K mole

     P in torr*
     V in liters
     n in moles
     T in ^{o}K
} $R = 62.36$ liter·torr/^{o}K mole

For solids and liquids, the volume of a given sample gives one an idea of the amount of matter present in the sample. In other words, the volume of a solid or liquid depends minimally on the temperature and pressure. For a sample of gas, however, the volume occupied by a given amount of matter depends entirely on the temperature and pressure. Because the volume of gases is so variable, it is often of interest to know what the volume of a gas sample would be under a standardized set of conditions. These conditions of standard temperature and pressure (STP) are as follows:

$$T = 0^{o}C = 273.15^{o}K$$
$$P = 1 \text{ atm} = 760 \text{ torr}$$

The conversion of the volume of a gas sample to the volume at standard conditions is easily programmed, as in Example 8.1.

Example 8.1

Program to convert ideal gas volumes to STP (necessary information: $P_{standard} = 1$ atm, $T_{standard} = 273.15^{o}$K)

*NOTE: 1 torr = 1 mm Hg.

```
95  REM PROGRAM TO CONVERT IDEAL GAS VOLUMES TO STP
100 PRINT "ENTER INITIAL VOLUME (LITERS)";
110 INPUT V1
120 PRINT "ENTER INITIAL PRESSURE (ATM),TEMPERATURE(K)";
130 INPUT P,T
140 LET V2=V1*(P/1)*(273.15/T)
150 PRINT "VOLUME AT STP =";V2;"LITERS"
160 END
RUN
ENTER INITIAL VOLUME (LITERS)?22.4
ENTER INITIAL PRESSURE (ATM),TEMPERATURE(K)?4,273.15
VOLUME AT STP = 89.6      LITERS
DONE
RUN
ENTER INITIAL VOLUME (LITERS)?5
ENTER INITIAL PRESSURE (ATM),TEMPERATURE(K)?.55,100
VOLUME AT STP = 7.51163    LITERS
DONE
```

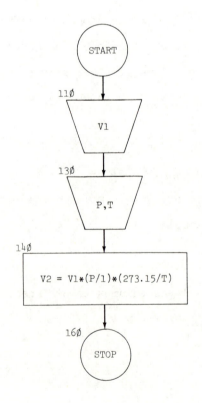

Figure 8.1

The actual conversion is performed in line 140. The standard temperature appears in the numerator in that line, while the standard pressure appears in the denominator. (Why?)

Problem 8.1

Modify the program in Example 8.1 to accept temperature in either centigrade or Kelvin and pressure in either atmospheres or torr (mm of mercury). (One atmosphere = 760 torr.)

Use your modified program to find the volume at STP of each of the following gas samples:

(a) 1 liter of Cl_2 at $0°C$ and 760 torr;

(b) 1 liter of Cl_2 at $100°C$ and 355.2 torr;

(c) 0.326 liter of Ar at $95°K$ and 0.01 atmosphere

Problem 8.2

Write a program to accept as input <u>any three</u> of the variables P, V, n, T and to calculate the value of the fourth variable (for an ideal gas). (Hint: use a "flag" to tell the program which variable is to be calculated. The flag here should be some unique input value that could not possibly be a real value for any of the variables. Because none of the four variables involved here can be negative, a negative number as an input flag would be appropriate.) Test your program in all modes of operation.

8.2 Computer Simulation and Graphic Output*

We shall now digress to develop a computer technique that will aid us greatly in the study of gases.

Most of our example programs so far have been of the "number crunching" type (see Introduction). It is now appropriate to explore the "simulation" potential of the computer. In this kind of application, the computer mimics the behavior of a physical system, usually by solving the equations that comprise a "mathematical model" of the system. An example is the use of computers to simulate automobile suspension systems so that engineers can optimize design features without building numerous costly prototypes.

*NOTE: For a fuller discussion of teletype graphics, see <u>Journal</u> <u>of</u> <u>Chemical</u> <u>Education</u>, 49: 357-361 (1972).

The use of the computer as a simulation device is greatly enhanced when the output is presented in graphic form rather than as tables of numbers. There are a number of computer output devices that are specially suited to graphic display, including on-line plotters and cathode-ray-tube terminals. With suitable programming, however, even the teletype terminal can be made to produce useful graphic output.

It is useful to identify two different types of plotting or graphing situation:

1   The data arises from an analytic function $y = f(x)$.

Such a case would be

$$P = nRT/V$$

where n and T are fixed and P is to be plotted as a function of V.

2   The data is given as a table of discrete data pairs. An example of this situation would be a table of experimentally obtained pressure-volume data for 1 mole of a certain gas at a particular temperature.

Example 8.2 incorporates a teletype plotting routine that you will be able to use for many simulation applications.

Example 8.2

General plotting routine for analytic functions: application to ideal gas pressure versus volume plot.

```
7∅  REM PROGRAM TO PLOT P VS. V FOR IDEAL GAS
8∅  PRINT "ENTER TEMPERATURE (KELVIN)";
85  INPUT T
88  REM LIMITS ON INDEP. VARIABLE
9∅  PRINT "ENTER XMIN";
92  INPUT X2
94  PRINT "ENTER XMAX";
96  INPUT X3
98  PRINT "ENTER XINCR";
1∅∅ INPUT X1
1∅2 PRINT
148 REM INTEGERIZING FUNCTIONS
15∅ DEF FNI(X)=INT(X/X1+.5)
```

```
160   DEF FNJ(Y)=INT((Y-Y2)/D+.5)+1

198   REM IDEAL GAS FUNCTION

200   DEF FNB(X)=.08205*T/X

300   PRINT "AUTOMATIC SCALING";

310   DIM A$[3]

312   INPUT A$

314   IF A$="NO" THEN 880

800   REM AUTO SCALING SECTION

805   LET Y=FNB(X2)

810   LET C1=C2=Y

815   FOR X=X2 TO X3 STEP X1

820   LET Y=FNB(X)

825   IF C1 <= Y THEN 835

830   LET C1=Y

835   IF C2 >= Y THEN 845

840   LET C2=Y

845   NEXT X

850   PRINT "YMIN= ";C1

855   PRINT "YMAX= ";C2

860   LET Y2=C1

865   LET Y3=C2

870   GOTO 900

875   REM LIMITS ON DEP. VARIABLE

880   PRINT "ENTER YMIN";

885   INPUT Y2

890   PRINT "ENTER YMAX";

895   INPUT Y3

900   LET D=(Y3-Y2)/70

905   PRINT "YINCR= ";D

910   GOSUB 9000

1000  REM PLOT GENERATOR

1005  FOR X=X2 TO X3 STEP X1

1010  LET Y=FNB(X)
```

```
1020 LET Y=FNJ(Y)

1030 PRINT ".";

1040 IF Y>71 THEN 1070

1050 IF Y<1 THEN 1070

1058 REM PRINTS POINT

1060 PRINT TAB(Y);"*";

1070 PRINT

1080 NEXT X

2000 GOSUB 9000

8997 REM AXES SUBROUTINE

8999 STOP

9000 PRINT " ";

9010 FOR I=1 TO 71

9020 PRINT ".";

9030 NEXT I

9040 PRINT

9050 RETURN

9999 END
```

Lines 88 to 9999 comprise a general-purpose plotting routine for functions which can be expressed in the form y = f(x). The function is established in the DEF FNB statement in line 200, with X treated as the independent variable. (Bear in mind that the actual variable name used in the DEF statement is arbitrary, because it represents a dummy variable.) Several successive DEF statements could be inserted here if necessary to construct a complicated function. In the present case, the function P = RT/V is defined by the statement DEF FNB(X) = .08205*T/X, because V is to be the independent variable. Note that the plots come out sideways.

This program gives the option of having the computer find appropriate limits for the dependent variable (YMIN, YMAX). In general, if a series of plots are to be compared, the first one should be run with automatic scaling, and then the same limits (YMIN, YMAX) should be INPUT manually for the rest of the plots.

You should make a tape of this routine for future use. Considerable time will be saved if your instructor stores this routine in the public or community library of your computer.

```
RUN
ENTER TEMPERATURE (KELVIN)?300
ENTER XMIN?.08
ENTER XMAX?.68
ENTER XINCR?.02
AUTOMATIC SCALING?YES
YMIN=  36.1985
YMAX=  307.687
YINCR=  3.87841
```

```
DONE
```

Example 8.3 makes use of a plotting routine for tabulated data pairs, the second type of plotting situation.

Example 8.3

General plotting routine for discrete data points: application to pressure versus volume data for CO_2 at $300^{\circ}K$

```
8Ø   REM GENERAL PLOTTING ROUTINE FOR DISCRETE POINTS
85   DIM X[1ØØ],Y[1ØØ]
88   REM LIMITS ON INDEP. VARIABLE
9Ø   PRINT "ENTER XMIN";
92   INPUT X2
94   PRINT "ENTER XMAX";
96   INPUT X3
98   PRINT "ENTER XINCR";
1ØØ  INPUT X1
1Ø2  PRINT
1Ø5  PRINT "ENTER YMIN";
11Ø  INPUT Y2
115  PRINT "ENTER YMAX";
12Ø  INPUT Y3
125  LET X8=INT(X2/X1)
13Ø  LET X9=INT(X3/X1+.5)
14Ø  LET D=(Y3-Y2)/7Ø
145  PRINT "YINCR= ";D
148  REM INTEGERIZING FUNCTIONS
15Ø  DEF FNI(X)=INT(X/X1+.5)
16Ø  DEF FNJ(Y)=INT((Y-Y2)/D+.5)+1
2ØØ  PRINT "NUMBER OF POINTS";
21Ø  INPUT N
218  REM SCALES AND INTEGERIZES THE DATA POINTS
22Ø  FOR I=1 TO N
23Ø  READ X[I],Y[I]
24Ø  LET X[I]=FNI(X[I])
```

```
25Ø   LET Y[I]=FNJ(Y[I])

26Ø   NEXT I

27Ø   GOSUB 9ØØØ

1ØØØ  REM PLOT GENERATOR

1ØØ5  FOR X=X8 TO X9

1Ø1Ø  PRINT ".";

1Ø2Ø  FOR I=1 TO N

1Ø3Ø  IF X[I]#X THEN 1Ø7Ø

1Ø35  IF Y[I]>71 THEN 1Ø7Ø

1Ø36  IF Y[I]< 1 THEN 1Ø7Ø

1Ø38  REM PRINTS POINT

1Ø4Ø  PRINT TAB(Y[I]),"*";

1Ø48  REM THERE IS A CTRL O INSIDE THE QUOTES IN LINE 1Ø5Ø

1Ø5Ø  PRINT "";

1Ø7Ø  NEXT I

1Ø8Ø  PRINT

1Ø9Ø  NEXT X

1Ø95  GOSUB 9ØØØ

11ØØ  REM DATA STATEMENTS CONTAIN X,Y PAIRS

11Ø1  DATA .Ø8,99,.Ø88,9Ø.3,.1Ø5,69.5

11Ø2  DATA .13,69.6,.155,69.8,.18,68.5

11Ø3  DATA .2Ø5,68.2,.23,63.6,.33,52.7

11Ø4  DATA .43,44.1,.53,37.7,.63,32.9

8997  REM AXES SUBROUTINE

8999  STOP

9ØØØ  PRINT " ";

9Ø1Ø  FOR I=1 TO 71

9Ø2Ø  PRINT ".";

9Ø3Ø  NEXT I

9Ø4Ø  PRINT

9Ø5Ø  RETURN

9999  END
```

This routine will plot up to 100 points, whose coordinates are entered as (X,Y) pairs in the DATA statements. The data shown in lines 1101 to 1104 represent volume, pressure data for 1 mole of CO_2 at $300^{\circ}K$. The volumes are in liters and the pressures are in atmospheres.

A tape should be made of this routine, or the routine should be stored in the computer.

```
RUN
ENTER XMIN?.08
ENTER XMAX?.68
ENTER XINCR?.02
ENTER YMIN?32
ENTER YMAX?100
YINCR=  .971429
NUMBER OF POINTS?12
```

```
DONE
```

Problem 8.3

Use the program in Example 8.2 to make pressure versus volume plots for one mole of an ideal gas at the following temperatures:

$$300^{\circ}K \quad 500^{\circ}K \quad 700^{\circ}K$$

Run the first plot with automatic scaling; then run the others using the same pressure limits as the first plot. A volume range of 1 liter to 10 liters in increments of 0.25 liters would be appropriate.

Problem 8.4

Run the same comparison as in problem 8.3, but use automatic scaling on the high temperature ($700^{\circ}K$) plot; then use the resulting pressure limits to make the $300^{\circ}K$ and $500^{\circ}K$ plots. How does this set of graphs differ from the set obtained in the previous problem?

Problem 8.5

Modify the program in Example 8.2 to make volume versus temperature plots for 1 mole of an ideal gas held at a constant pressure of 1 atm. Run a plot with a temperature range of $100^{\circ}K$ to $600^{\circ}K$ in increments of $10^{\circ}K$; use volume limits of 0 liters and 50 liters.

How would you describe the resulting plot? Is every minute feature of the plot genuinely related to the behavior of the gas? What characteristic of the teletype terminal might introduce distortion into a plot?

Using a straightedge, extrapolate the plot back to a volume of 0 liters. What temperature on the plot does this volume correspond to? Is this what you would expect?

Problem 8.6

Modify the program in Example 8.2 to make pressure versus temperature plots for 1 mole of an ideal gas confined in a constant volume of 1.1 liter. Run a plot with a temperature range of $100^{\circ}K$ to $600^{\circ}K$ in increments of $10^{\circ}K$; use pressure limits of 0 atm and 50 atm.

How would you describe the resulting plot? Is every minute feature of the plot genuinely related to the behavior of the gas?

Using a straightedge, extrapolate the plot back to a pressure of 0 atm. What temperature on the plot does this pressure correspond to? Is this what you would expect?

8.3 Real Gases

It appears from the sample runs in Examples 8.2 and 8.3 that the ideal gas model, at least under the conditions specified in those examples, does not agree perfectly with the experimental data for a real gas. This is not surprising in view of the two very unrealistic hypothetical attributes of the ideal gas mentioned earlier in this chapter.

There are, however, gas equations that attempt to incorporate both the finite molecular volume and the intermolecular attraction present in real gases. One of these equations is the van der Waals equation,

$$(P + n^2 a/V^2)(V - nb) = nRT$$

or, for 1 mole of gas,

$$(P + a/V^2)(V - b) = RT$$

In the van der Waals equation, a and b are called the "van der Waals constants" and have different values for each different real gas.

If V and T are known for 1 mole of a particular gas, then calculation of P from the van der Waals equation is straightforward:

$$P = RT/(V - b) - a/V^2$$

If, on the other hand, P and T are given, finding V is more difficult, because the van der Waals equation is cubic in V.

Values of the van der Waals constants for some common gases are given in Table 8.1.

Table 8.1 Van der Waals constants for some common gases.[*]

Gas	a (liter^{2}atm/mole^2)	b (liter/mole)
Argon	1.35	3.23×10^{-2}
Carbon dioxide	3.60	4.28×10^{-2}
Chlorine	6.50	5.64×10^{-2}
Hydrogen	0.245	2.67×10^{-2}
Oxygen	1.36	3.19×10^{-2}
Sulfur dioxide	6.72	5.65×10^{-2}
Water	5.46	3.30×10^{-2}

*The van der Waals constants are somewhat temperature dependent, so these values should be considered to be approximate.

Problem 8.7

Write a program to compute the pressure of 1 mole each of an ideal gas and a van der Waals gas as a function of increasing volume, given the temperature as an INPUT datum. Have your program present the output in tabular form:

P(IDEAL)   P(V.D.W.)   VOLUME

Use your program to compare the ideal gas and van der Waals gas pressures in the volume range 0.08 liters to 0.40 liters using volume increments of about 0.01 liters. Run this comparison for CO_2 first at $700^\circ K$ and then at $304^\circ K$. What do you notice about the comparative behavior of the two gas models in each case?

Because an ideal gas has no intermolecular attractions, such a gas (if it existed) would never liquefy. Real gases, of course, can be liquefied by compression and cooling; but for every real gas, there is a temperature above which the gas cannot be liquefied, regardless of compression. This temperature is called the "critical temperature". For CO_2, the critical temperature is $304.1^\circ K$. Can you now begin to explain the comparative behavior of the real versus ideal gas models for CO_2 at the two temperatures you investigated?

Problem 8.8

Use the program in Example 8.2 to make P versus V plots for 1 mole of a van der Waals gas (use the van der Waals constants for CO_2) under the following sets of conditions:

(a) $T=320^\circ K$ $V=0.08$ liters to 0.40 liters in increments of 0.005 liters;

(b) $T=304^\circ K$ same volume limits;

(c) $T=290^\circ K$ same volume limits.

Problem 8.9

Show that the van der Waals equation is cubic in V; that is, rearrange the equation to the form

$$AV^3 + BV^2 + CV + D = 0$$

(Note that this task is better done by a human than by a computer.)

If your computer system has a library program for finding roots of polynomials, modify that program to prepare a table of V versus P for one mole of a van der Vaals gas. Run the program with a temperature of $647^\circ K$ using the van der Waals constants for water. Hand plot the resulting values. Since the equation is cubic, there will be three solutions for each value of P; however, only real solutions (not imaginary or complex) can be physically meaningful.

If your computer does not have the necessary library program, the program in Chapter 10, Example 10.2 can be used if the volume does not exceed 50 liters.

8.4 Partial Pressure

So far, we have considered only pure gases. If instead we have a mixture of gases confined in a vessel of volume V, then the <u>total</u> pressure in the vessel is the sum of the pressures that each individual gas would exert if it alone occupied the vessel. That is,

$$P_{total} = P_1 + P_2 + P_3 + \cdots$$

The pressures of the individual gases (P_1, P_2, P_3, etc.) are called the "partial pressures" of those gases.

The mole fraction X_i of each gas

$$X_i = \frac{\text{moles of gas } i}{\text{total moles of all gases present}}$$

is related to the partial pressure as

$$X_i = \frac{P_i}{P_{total}}$$

This can be seen by observing that

$$n_i = \frac{P_i V}{RT}$$

and

$$n_{total} = \sum_i n_i = \sum_i \frac{P_i V}{RT} = \frac{V}{RT} \sum_i P_i = \frac{V}{RT} P_{total}$$

Then,

$$X_i = \frac{n_i}{n_{total}} = \frac{\frac{P_i V}{RT}}{\frac{V}{RT} P_{total}} = \frac{P_i}{P_{total}}$$

If there is a liquid in the container as well as the gases, another factor comes into play. When a liquid substance is confined in a container that is larger than the volume of the liquid itself, the liquid will give off vapor until an equilibrium is established between the liquid and the gaseous form of the substance. At any particular temperature, the gas phase of the substance will have a certain pressure that is characteristic of that liquid at that temperature. This characteristic pressure is called the "vapor pressure" of the liquid. The vapor pressure of a liquid increases with increasing temperature; when the vapor pressure equals atmospheric pressure, the liquid boils. Now, when a liquid is present in a vessel containing other gases, the vapor of the liquid acts like the other gases in contributing to the total pressure in the container. The partial pressure contributed by the vapor is equal to the vapor pressure of the liquid.

Problem 8.10

Write a program to compute the partial pressures of all gases in a mixture given the mole fraction of each gas and the total pressure of the mixture. Use the program to calculate the partial pressures of the following atmospheric constituents at atmospheric pressure (760 torr).

Gas	Mole fraction in the atmosphere
N_2	0.7808
O_2	0.2095
Ar	0.0093
CO_2	0.0003
Ne	0.00002
He	0.000005
CH_4	0.000002

Problem 8.11

In the laboratory, gases are often collected over water. In such cases, the measured gas pressure includes the vapor pressure of the water. Therefore, in correcting the volume of the collected gas to STP, the vapor pressure of the water must first be subtracted from the measured pressure.

Modify your program from Problem 8.1 in order to correct to STP the volume of a gas sample collected over water. Include an information retrieval feature so that the program will apply the appropriate vapor pressure correction based on the INPUT temperature. Use the data from the following vapor pressure table for water.

Table 8.2 Vapor pressure of water.

T, $^{\circ}$C	Vapor pressure, torr	T, $^{\circ}$C	Vapor pressure, torr
0	4.58	26	25.2
1	4.93	27	26.7
2	5.29	28	28.3
3	5.68	29	30.0
4	6.10	30	31.8
5	6.54	31	33.7
6	7.01	32	35.7
7	7.51	33	37.7
8	8.04	34	39.9
9	8.61	35	42.2
10	9.21	36	44.6
11	9.84	37	47.1
12	10.5	38	49.7
13	11.2	39	52.4
14	12.0	40	55.3
15	12.8	41	58.3
16	13.6	42	61.5
17	14.5	43	64.8
18	15.5	44	68.3
19	16.5	45	71.9
20	17.5	46	75.6
21	18.6	47	79.6
22	19.8	48	83.7
23	21.1	49	88.0
24	22.4	50	92.5
25	23.8		

Use your program to correct to STP the volumes of the following samples of gases collected over water:

(a) 1 liter of He at 0°C and 760 torr;

(b) 0.53 liter of O_2 at 23°C and 160 torr;

(c) 3 liters of N_2 at 45°C and 623 torr;

(d) 1.5 liter of H_2 at 15°C and 150 torr;

(e) 1 liter of Xe at 49°C and 100 torr.

8.5 Kinetic Theory of Gases

One can measure and discuss the pressure, volume, and temperature of a gas sample without ever considering the molecular makeup of the gas. Such an approach is called "macroscopic"; and, while it can describe gas behavior, the macroscopic approach offers no explanation as to why gases behave as they do.

The "why" is to be found in the realm of (sub)microscopic considerations, and the kinetic theory of gases proposes a microscopic model for the nature of gases that attempts to account for their observable behavior. The kinetic theory assumes:

1 that a gas is composed of molecules which are essentially point masses;

2 that these molecules collide elastically but otherwise exert no forces on each other;

   (Do you recognize the definition of an ideal gas implicit in these two assumptions?)

3 that the individual molecules are not restricted as to possible values for their kinetic energies.

From these simple postulates, plus the association of the average kinetic energy of a gas sample with its temperature

$$\bar{E} = \frac{3}{2} RT$$

much of the macroscopic, observable behavior of gases can be predicted.

The application of statistical concepts to the kinetic molecular model results in an even more detailed picture of what goes on "inside" a sample of gas. We shall use the computer to explore three interesting manifestations of this model.

8.6 Molecular Speeds

A consequence of the statistical treatment of the kinetic molecular model is the Maxwell-Boltzmann relation. This tells the actual distribution of molecular speeds in a given gas sample. The somewhat intimidating formula is

$$p_c dc = 4\pi \left(\frac{m}{2\pi kT}\right)^{3/2} e^{-\left(\frac{m}{2kT} c^2\right)} c^2 dc$$

The meaning of this expression is actually quite simple.

If you have a large collection of molecules (so that statistics is meaningful), each with mass m (in grams), at temperature T (^{o}K), then the fraction of the molecules having speeds between c (in cm/sec) and c + dc is given by plugging the chosen values of m, T, and c into the formula. The constant k is just the gas constant R divided by Avogadro's number. In the units appropriate here, k = 1.38 × 10^{-16} erg/^{o}K.

It is of interest to graph p_c versus c. We can quickly find the shape of such a plot by using the plotting routine in Example 8.2. This is done in Example 8.4.

Example 8.4

Plot of Maxwell-Boltzmann distribution of molecular speeds: run for neon at 300°K.

```
50   PRINT "ENTER MASS OF ATOM OR MOLECULE (AMU)";
52   INPUT M1
54   LET M=M1/6.024E+23
60   PRINT "ENTER TEMPERATURE (K)";
62   INPUT T
65   K=1.38E-16
67   P1=3.14159
70   REM USE LINES 88-9999 VERBATIM FROM EXAMPLE 8.2
200  DEF FNB(C)=4*P1*(M/2/P1/K/T)↑1.5*C↑2*EXP(-(M/2/K/T)*C↑2)
RUN
ENTER MASS OF ATOM OR MOLECULE (AMU)?20.183
ENTER TEMPERATURE (K)?300
ENTER XMIN?0
ENTER XMAX?2.E05
ENTER XINCR?5.E03
AUTOMATIC SCALING?YES
YMIN=   0
YMAX=   1.66992E-05
YINCR=   2.38560E-07
```

(Run continued on next page)

DONE

This plot was run for neon (atomic weight = 20.183 amu) at 300°K. The x axis (representing speed) was scaled from 0 to 200,000 cm/sec in increments of 5,000 cm/ sec. The computer automatically scaled the y axis. It is important to realize that the Y value of this plot at any point does not represent a fraction or a probability.* Only over a finite range of speeds, Δc, can such a fraction be determined; the fraction is the <u>area</u> under the curve at c, over the interval Δc.

Problem 8.12

Use the program of Example 8.4 to plot the speed distribution (300°K) for helium. You may have to do some experimentation to determine the best limit for the x axis. Let the computer do the y scaling.

Then, using the same scaling, run the 300°K speed distributions for neon, argon, krypton, xenon, and radon. What conclusions can you draw from this series of plots?

Problem 8.13

Use the program of Example 8.4 to plot the speed distribution for Cl_2 at each of several different temperatures: say, 100°K, 273°K, 400°K, 600°K, 800°K. Run all plots at the same scale. What conclusions can you draw from this series of plots?

Problem 8.14

Chemical reactions involving the reaction of two or more molecules require that the molecules collide with each other. Suppose that, for a certain gas phase chemical reaction to occur, the colliding molecules need to have a speed of at least 5×10^4 cm/sec. The relative rate of this reaction at various temperatures would be proportional to the number of molecules moving at this speed or faster. This, in turn, would be proportional to the area under the Maxwell-Boltzmann speed distribution curve between $c = 5 \times 10^4$ cm/sec and $c = \infty$ cm/sec.

From your plots obtained in Problem 8.13 for Cl_2, estimate the reaction rate at each temperature relative to the reaction rate at 273°K for a simple gas phase collisional reaction involving Cl_2. (You can do this by cutting out and weighing the relevant portion of each plot. If your computer library has a program for numerical integration, you could do the integration on the computer. It would be interesting to compare the results of both methods.)

*NOTE: The actual value of the curve <u>at a point</u> is called a "probability density".

8.7 Brownian Motion and Monte Carlo Simulation

An observation that provides convincing support for the kinetic molecular theory of gases is that of Brownian motion. A very small particle suspended in a gas or liquid can be seen to undergo a random zigzag motion, ostensibly as it receives randomly directed impacts from the molecules of the gas or liquid.

Would it not be interesting to attempt to simulate Brownian motion on the computer, using the kinetic theory as a model? To try this brings us to an important point regarding computer simulation. Our previous simulation of ideal gas behavior (Example 8.2) was of the type known as "deterministic" simulation. The behavior of the system in a deterministic model is completely described by a set of analytic equations. For processes in which chance, or probability, is an explicit factor, however, one uses a different type of model, which itself partakes of the probabilistic character of the system to be simulated. Such a simulation is called a "stochastic" or "Monte Carlo" simulation.

In the following example, the Monte Carlo method is used to simulate the path of a particle undergoing Brownian motion in two dimensions. We will assume for simplicity that the displacement of the particle following collision with a molecule is proportional to the speed of the colliding molecule. We will also ignore the possibility of glancing collisions and multiple collisions.

Since the molecules in the gas are postulated to be moving about at random, the displacement of the particle resulting from an impact will also be random. We can simulate this effect by having the computer generate two random numbers (X1 and Y1) that can be added to the starting x and y coordinates of the particle. However, while the directions of the impinging molecules are truly random, the speeds of the molecules are not random over the whole range of possible speeds but are distributed according to the Maxwell-Boltzmann distribution.[*] Some speeds are more probable than others, whereas a random distribution is one in which all values are equally probable. Thus, we must impose this definite distribution on our random process. This is achieved by means of a trick common in Monte Carlo simulations.

[*]NOTE: Since we are restricting the motion to two dimensions, we must use a slightly different form of the Maxwell-Boltzmann relation:

$$p_c \, dc \;=\; 2\pi \left(\frac{m}{2\pi kT}\right) e^{-\left(\frac{m}{2kT} \, c^2\right)} c \, dc$$

From the first two random numbers chosen, we can calculate the displacement that would result from our proposed collision:

$$D = SQR(X1 \uparrow 2 + Y1 \uparrow 2)$$

We assume that the speed of the colliding molecule is proportional to the displacement caused by the collision. From the two-dimensional Maxwell-Boltzmann formula, we can calculate the probability of a molecule having a speed within an arbitrary range of the proposed speed of the colliding molecule. We now can have the computer generate a third random number. If the value of this third random number is less than the probability of occurrence for the particular speed range in question, then we "accept" the collision and apply the displacement to the position coordinates of the Brownian particle. If, on the other hand, the third random number is greater than the actual probability for the particular speed range, then we "reject" this particular displacement and generate two new displacement random numbers. For example, if the particular speed range has a probability of 0.05, then any value of the third random number from 0 to 0.05 would cause us to accept the displacement. Values in the range 0.05 to 1 would lead us to reject the trial collision. The larger the probability for the speed range associated with a trial collision, the more likely we are to accept the collision. Thus, after a large number of colli- sions, the number of displacements of various magnitudes will be proportional to the actual speed distribution of the gas molecules, and the path of the particle receiving the impacts should be that of a true, two-dimensional Brownian motion.

The qualification of large number of trials is crucial. Because of the sta- tistical nature of Monte Carlo simulations, they work well only when run for a large number of trials. But this limitation actually parallels the true behavior of the system being simulated. The behavior of a gas sample containing only five molecules would be erratic and unpredictable. For a real sample containing, say, 10^{23} mole- cules, the number of collisions is so large that the random motions of individual molecules average out for the whole assemblage, and it becomes meaningful to talk of average speed, of average energy, and of temperature as being representative of the sample as a whole.

Example 8.5

Program for Monte Carlo simulation of two-dimenstional Brownian motion.

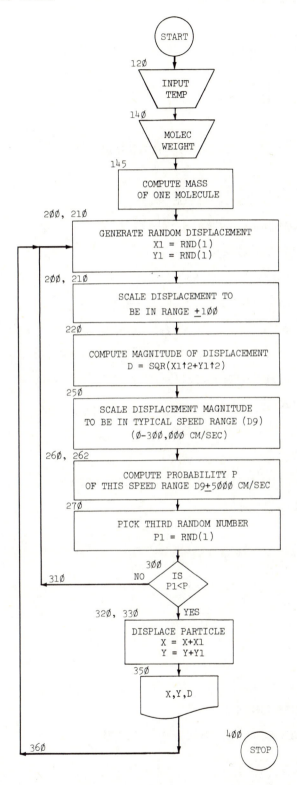

Figure 8.2 Monte Carlo simulation of Brownian motion.

```
100   REM BROWNIAN MOTION SIMULATION
110   PRINT "ENTER TEMPERATURE (K)";
120   INPUT T
130   PRINT "ENTER MASS OF ATOM OR MOLECULE (AMU)";
140   INPUT M1
145   LET M=M1/6.024E+23
150   LET X=Y=0
160   PRINT "X","Y","DISPLACEMENT MAGNITUDE"
170   PRINT 0,0
198   REM PICK TRIAL NEW POSITION
200   LET X1=-100+RND(1)*200
210   LET Y1=-100+RND(1)*200
220   LET D=SQR(X1↑2+Y1↑2)
248   REM SCALE DISPLACEMENT TO SPEED RANGE
250   LET D9=D*3000
258   REM COMPUTE PROBABILITY OF SPEED IN THIS RANGE
260   LET P9=M/1.38E-16/T*D9*EXP(-(M/2/1.38E-16/T)*D9↑2)
262   LET P=P9*10000
268   REM PICK RANDOM PROBABILITY
270   LET P1=RND(1)
298   REM SECTION TO ACCEPT OR REJECT DISPLACEMENT
300   IF P1<P THEN 320
310   GOTO 200
320   LET X=X+X1
330   LET Y=Y+Y1
350   PRINT X,Y,D
360   GOTO 200
400   END
RUN
ENTER TEMPERATURE (K)?300
ENTER MASS OF ATOM OR MOLECULE (AMU)?20.183
```

(Run continued on next page)

X	Y	DISPLACEMENT MAGNITUDE
0	0	
-11.3636	.422241	11.3714
-9.84758	-16.7544	17.2434
-20.3306	-36.0271	21.9393
-5.48915	-41.2902	15.747
-5.71855	-23.2055	18.0861
-24.6541	-8.14847	24.1924
-15.4827	-12.3023	10.0683
-21.3467	.994812	14.5327
-20.4641	16.129	15.1599
-.782806	3.34746	23.4675

STOP

Lines 100 to 170 of this program simply INPUT the necessary data and label the output list. Lines 200 and 210 are of some interest. They utilize the RND function to choose the trial coordinate displacements. The argument of the RND function is a dummy, so we arbitrarily use RND(1). The random numbers generated by the computer are always between 0 and 1, but this particular range may not always be convenient. In our case, to facilitate graphing, we might like the coordinate displacements to be in the range -100 to +100. To convert the random number produced by the RND function to a random number in the range -100 to +100, lines 200 and 210 incorporate an "algorithm" or procedure to effect the conversion. The algorithm here says, "Multiply the random number by 200 and subtract 100." Satisfy yourself that this algorithm has the desired effect.

In line 220, we calculate the magnitude of the displacement associated with the coordinate displacements, X1 and Y1, using the Pythagorean formula. We have assumed for this problem that the displacements of the Brownian particle are proportional to the speed of the impinging molecule. To convert the displacement D to a number representing a molecular speed, we build an algorithm to take a number in the range 0 to, say, 100 and to convert that number to a value in the (somewhat arbitrary) range 0 to 300,000. This very simple algorithm is embodied in line 250. The upper limit (300,000) was chosen by examining plots from the Maxwell-Boltzmann program (Example 8.4).

In line 260, we use the Maxwell-Boltzmann equation to evaluate the probability density for the molecular speed corresponding to our trial displacement. Recalling that the actual probability of a speed range is the _area_ under the Maxwell-Boltzmann curve in that speed range, we choose an arbitrary speed interval (10,000 cm/sec) and use a rectangle to approximate the area under the curve; this is done in line 262. The rectangle, of course, is the crudest possible approximation to the area under the curve, and the program could be refined by using a more sophisticated integration procedure.

We now choose the third random number, P1, in line 270. In line 300, we compare P1 with the actual probability for our speed range. If the collision is accepted, the coordinates (X,Y) of the Brownian particle are modified by the addition of X1 and Y1. If the collision is rejected, we go back to line 200 and start again with a new trial collision. Notice that this program does not end by itself. To terminate execution, you must depress the BREAK button on your terminal.

The graph below shows the path of a particle undergoing Brownian motion as simulated by this program.

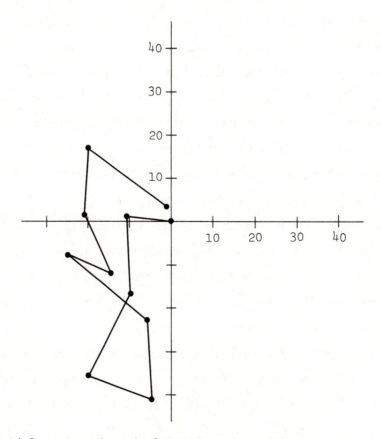

Figure 8.3 Particle undergoing simulated Brownian motion.

A word of caution is in order regarding the RND function. Each time this function is called within the execution of a program, a new random number between 0 and 1 is produced by the computer. One might expect that if the program is RUN again, a new series of random numbers would be obtained. However, with some computers exactly the same sequence of random numbers is produced if the program is re-RUN. If your computer is like that, you should build in a randomizing loop at the beginning of the program. This is just a sequence which READs through the random number table to an arbitrary point so that you don't always start in the same place. An example of such a loop would be:

```
5Ø PRINT "ENTER A NUMBER FROM 1 TO 1ØØØ";
52 INPUT Z
54 FOR I=1 TO Z
56 LET R=RND(1)
58 NEXT I
```

Problem 8.15

Using the Brownian motion simulation in Example 8.5, run several simulations of a particle's path at each of the following temperatures:

$$100^{\circ}K \qquad 500^{\circ}K \qquad 1,000^{\circ}K$$

Allow the program to run for at least 25 collisions each time. Manually plot the particle's journey for one run at each temperature. Do you notice a trend?

Compare three separate journeys at the same temperature.

Problem 8.16

The importance of large numbers of trials for Monte Carlo simulation has been stressed. Modify the program in Example 8.5 to keep track of the number of trials that are made for each collision that is actually accepted. Have the number of trials printed as part of the program output. Add up how many trials are made by the computer for a journey of, say, 25 collisions.

8.8 Graham's Law of Diffusion

One consequence of the kinetic molecular theory is the association of the temperature of a gas with the average kinetic energy of the molecules. Specifically,

$$\overline{E} = 3/2 \ RT$$

Thus, at a given temperature ($^\circ$K), the mean molecular kinetic energy of all (ideal) gases would be the same. Thus, if we consider two different gases with gram-molecular weights M_1 and M_2 (or Nm_1 and Nm_2 where N is Avogadro's number and m_1 and m_2 are the masses of one molecule of each gas), then

$$\bar{E}_1 = \bar{E}_2$$

$$1/2\ m_1\ v_1^2 = 1/2\ m_2\ v_2^2$$

$$\frac{v_1^2}{v_2^2} = \frac{m_2}{m_1}$$

and

$$\frac{v_1}{v_2} = \sqrt{\frac{m_2}{m_1}} = \sqrt{\frac{Nm_2}{Nm_1}} = \sqrt{\frac{M_2}{M_1}}$$

This last expression is Graham's Law of Diffusion. It simply shows that the relative average molecular velocities of two different gases are inversely proportional to the square roots of the molecular weights.

Problem 8.17

Write a program to compute the relative molecular velocities of any two gases, given their molecular weights. Use this program to calculate the velocities of the following gases relative to hydrogen (H_2):

$$HCl \qquad DCl \qquad NH_3 \qquad O_2 \qquad CO_2 \qquad N_2$$

Problem 8.18

Use the program from Problem 8.17 to calculate the diffusion rate of NH_3 relative to that of HCl. Test the result by placing a bottle of concentrated HCl and a bottle of concentrated aqueous NH_3 about 1 meter apart on a table in a place free from drafts. Remove the stoppers.

Where does the cloud of white NH_4Cl form first? Why?

9
Solutions

For a variety of reasons, a great preponderance of chemical reactions are carried out in solutions. Solutions provide a convenient way of insuring a homogeneous
reaction environment, of handling solids and gases, and of controlling relative
amounts of reactants.

9.1 Concentration

In using a solution as a vehicle for a reactant, it is often important to know
the amount of reactant present in a given amount of solution. This information is
given by the "concentration", which can be defined in various ways. The most commonly used measure of concentration is molarity; molarity is defined as the number
of moles of solute per liter of solution. The program in Example 9.1 is based on
this definition.

Example 9.1

The following is a program to calculate the molarity of a solution given the
mass and molecular weight of solute and the volume of solution.

```
9Ø    REM PROGRAM TO COMPUTE MOLARITY
1ØØ   PRINT "ENTER MOLECULAR WEIGHT OF SOLUTE";
11Ø   INPUT M
12Ø   PRINT "ENTER MASS OF SOLUTE (G)";
```

```
13∅    INPUT G

14∅    PRINT "ENTER VOLUME OF SOLUTION (ML)";

15∅    INPUT V

16∅    LET N=G/M

17∅    LET C=(N/V)*1∅∅∅

175    PRINT

18∅    PRINT "SOLUTION IS";C;"MOLAR"

19∅    END
RUN
ENTER MOLECULAR WEIGHT OF SOLUTE?46.∅7
ENTER MASS OF SOLUTE (G)?23.∅35
ENTER VOLUME OF SOLUTION (ML)?25∅
SOLUTION IS 2.           MOLAR
DONE
```

Here, line 160 calculates the number of moles, N. Line 170 calculates the molarity. Both these calculations could have been combined into one step, of course. The factor of 1,000 in line 170 converts the units from moles/ml to moles/liter.

In some situations, it is more useful to know the number of moles of solute per 1,000 grams of solvent. This measure of concentration is called "molality". The program in Example 9.1 can be modified as follows to calculate molality. Notice that it is also necessary to know the density of the solvent if we are not given the solvent mass itself.

Example 9.2

The following is a program to calculate the molality of a solution given the mass and molecular weight of solute and the volume and density of solvent.

```
9∅     REM PROGRAM TO COMPUTE MOLALITY

1∅∅    PRINT "ENTER MOLECULAR WEIGHT OF SOLUTE";

11∅    INPUT M

12∅    PRINT "ENTER MASS OF SOLUTE";

13∅    INPUT G

14∅    PRINT "ENTER VOLUME OF SOLVENT (ML)";

15∅    INPUT V
```

```
152   PRINT "ENTER DENSITY OF SOLVENT (G/ML)";

154   INPUT D

160   LET N=G/M

165   LET W=V*D

170   LET C=(N/W)*1000

175   PRINT

180   PRINT "SOLUTION IS";C;"MOLAL"

190   END

RUN

ENTER MOLECULAR WEIGHT OF SOLUTE?46.07

ENTER MASS OF SOLUTE?23.035

ENTER VOLUME OF SOLVENT (ML)?250

ENTER DENSITY OF SOLVENT (G/ML)?.791

SOLUTION IS 2.52844    MOLAL

DONE
```

Here, line 160 computes the number of moles of solute, as in Example 9.1. Line 165 computes the number of grams of solvent. Line 170 calculates the molality, and a factor of 1,000 is again necessary, this time to convert from moles/g to moles/kg.

Problem 9.1

Write a program to calculate the molarity of a mixture of two portions of a given kind of solution when the two portions may be of different volumes and may have different concentrations. (Hint: have the program first calculate the total number of moles of solute. Notice that if the molarity of one portion is zero, the process is just that of dilution.)

Use your program to compute the solute concentrations resulting from the following combinations:

Portion 1		Portion 2
25.6ml of 0.132 molar $CaCl_2$	+	12.2ml of 0.154 molar $CaCl_2$
1.00ml of 10.0 molar H_2SO_4	+	10.0ml of 1.00 molar H_2SO_4
53.2ml of 6.04 molar HCl	+	2.35ml of 6.04 molar HCl
2.36 l of 0.154 molar $LiNO_3$	+	2.00 l of pure H_2O

Problem 9.2

Write a program incorporating information retrieval features to calculate the molarity of a solution of any binary salt of the first 20 elements, given the formula, weight of salt used, and volume of solution.

Problem 9.3

Modify the program in Problem 9.2 to calculate the amount of salt needed to make a desired amount of solution of a desired concentration.

Problem 9.4

What additional information is needed to convert the molarity of a solution into the corresponding molality?

Prepare a program that accepts this necessary information and then performs the conversion from molarity to molality. Use your program to convert the following molarities of ethyl alcohol to molalities:

Molarity of C_2H_5OH	Density of solution (g/cc)
2.136	0.9839
4.237	0.9715
6.236	0.9574
8.164	0.9393
17.232	0.7939

9.2 Colligative Properties

When a material is used as a solvent, the addition of the solute causes some of the properties of the solvent to change. If the change in a property depends solely on the concentration of solute particles and not on their identity, then that particular property is called a "colligative" property. Commonly studied colligative properties are vapor pressure, boiling point elevation, freezing point depression, and osmotic pressure.

For example, the normal boiling point of water is $100^{\circ}C$. However, for each mole of particles dissolved in a kilogram of water, the boiling point increases by $0.512^{\circ}C$. The following program uses this information to prepare a table of boiling points of salt solutions of varying concentration.

Example 9.3

Program to prepare a table of boiling point versus concentration for NaCl solutions (necessary information: the molal boiling point elevation for water is $0.512^{\circ}C$; NaCl dissociates in water to give 2 moles of dissolved particles per mole of NaCl).

```
90    REM PROGRAM FOR BOILING POINT TABLE
100   PRINT "ENTER CONCENTRATION LIMITS IN MOLES OF NACL/KG OF WATER"
110   PRINT " LOWER LIMIT";
120   INPUT X2
130   PRINT " UPPER LIMIT";
140   INPUT X3
150   PRINT "CONCENTRATION INCREMENT";
160   INPUT X1
167   REM NEXT WE SUPPLY THE MOLAL B.P. ELEVATION FOR WATER
169   REM AND THE NUMBER OF DISSOCIABLE PARTICLES FOR NACL
170   READ E,N
172   DATA .512,2
178   PRINT
180   PRINT "MOLALITY","B.P. IN DEGREES C"
190   FOR I=X2 TO X3 STEP X1
200   LET T=100+I*N*E
210   PRINT I,T
220   NEXT I
310   END
RUN
ENTER CONCENTRATION LIMITS IN MOLES OF NACL/KG OF WATER
 LOWER LIMIT?0
 UPPER LIMIT?2
CONCENTRATION INCREMENT?.2
```

(Run continued on next page)

MOLALITY	B.P. IN DEGREES C
0	100
.2	100.205
.4	100.41
.6	100.614
.8	100.819
1	101.024
1.2	101.229
1.4	101.434
1.6	101.638
1.8	101.843
2	102.048

DONE

Here, the molal boiling point elevation is READ in as E; and the number of dissociable particles per formula unit of solute is READ in as N. This program as it appears here would apply to solutions of any binary salt in water.

Problem 9.5

   Modify the program in Example 9.3 to prepare a boiling point table for solutions of $CaBr_2$. This compound dissociates into 3 moles of particles per mole of dissolved $CaBr_2$.

Problem 9.6

   The molecular weight of an unknown material can be determined by dissolving a weighed amount of the material in a known amount of a suitable solvent and measuring the amount by which the solvent's freezing point is depressed. Camphor has a normal melting point of $178.4°C$ and a molal freezing point depression of $37.7°C$. That is, the freezing point drops by $37.7°C$ for each mole of solute particles per 1,000 g of camphor.

   Write a program to compute the molecular weight of unknown materials given the weight of unknown used, the weight of camphor used, and the observed freezing point of the mixture. With your program, calculate the molecular weight of each of the following materials using the data given:

Substance 1.   1.02 g of unknown dissolved in 25.1 g camphor;

melting point = $169.9^{\circ}C$

Substance 2.   0.873 g of unknown dissolved in 20.9 g camphor;

melting point = $166.1^{\circ}C$

Substance 3.   1.55 g of unknown dissolved in 28.0 g camphor;

melting point = $169.7^{\circ}C$

Substance 4.   2.10 g of unknown dissolved in 30.1 g camphor;

melting point = $134.7^{\circ}C$.

Problem 9.7

It is of interest to know how much antifreeze to add to an automobile radiator
to lower the freezing point of the contents to various temperatures for winter
driving. Create a program to prepare a table that tells the freezing point of the
contents of a 7-quart radiator for various amounts of added antifreeze, from 1
quart to 6 quarts in increments of 0.25 quart.

Necessary information: the density of water may be taken as 1 g/cc; the density of permanent antifreeze (ethylene glycol) is 1.11 g/cc; the molecular weight
of ethylene glycol is 62.1 g/mole; and the freezing point depression of water is
$1.86^{\circ}C$/mole of particles dissolved in 1 kg of water. (Do not forget the conversion between Fahrenheit and centigrade and between quarts and liters or cc's.
Remember also that the total capacity of the radiator is 7 quarts.)

Problem 9.8

Vapor pressure is a colligative property. The vapor pressure of a solvent in
a solution, if the solution is "ideal", is $P = XP_o$ where P_o is the vapor pressure
of the pure solvent at the given temperature and X is the mole fraction of the
solvent. The mole fraction, X, is given by

$$X_{solvent} = \frac{\text{moles solvent}}{\text{moles solvent + moles solute particles}}$$

The following data are the actual vapor pressures of a series of $CaCl_2$ solutions.

Weight percent $CaCl_2$	Vapor pressure at $25^{\circ}C$ (torr)
0.00	23.76
9.33	22.57
14.95	21.38
19.03	20.19
22.25	19.00

Write a program to convert these weight percentages to mole fractions of water. (Hint: assume 100 g of solution.)

Write a program to prepare a table of vapor pressure versus mole fraction of solvent for an ideal solution. P_o for water at $25^{\circ}C$ is 23.76 torr. Run your program and compare the results with the above data for $CaCl_2$ solutions. What can you conclude about the ideality of $CaCl_2$ solutions of various concentrations?

10
Equilibrium

The word "equilibrium" suggests balance. A chemical system is in equilibrium when all the measurable variables characterizing the system are in balance, that is, the variables are undergoing no change. Thus, in a system at equilibrium, the temperature, pressure, volume, and the concentrations of all chemical species present--all of these remain constant.

10.1 The Equilibrium Constant

A very important relationship exists among the concentrations of the various chemical species present in an equilibrium situation. If the chemical reaction leading to the equilibrium is

$$aA + bB + cC \rightleftharpoons rR + sS + tT$$

then it is true that

$$\frac{[R]^r[S]^s[T]^t}{[A]^a[B]^b[C]^c} = K$$

K is called the equilibrium constant. The equilibrium constant then is equal to the product of the reaction product concentrations, each raised to the power of the corresponding stoichiometric coefficient, divided by the product of the reactant concentrations raised to the power of the reactant coefficients.

For example, in the reaction

$$2SO_2 + O_2 \rightleftharpoons 2SO_3$$

the expression

$$\frac{[SO_3]^2}{[SO_2]^2[O_2]}$$

is equal to a constant. The value of this constant changes if the temperature is changed. But if the temperature is held constant, then the above expression will have the same value no matter how the various concentrations are changed. If more O_2 is introduced (thereby increasing $[O_2]$), then some SO_2 will react with the O_2 producing more SO_3. The resulting changes in $[SO_2]$ and $[SO_3]$ will balance the change in $[O_2]$, and the overall expression will retain its original constant value.

The equation for the equilibrium constant thus allows us to compute the concentration of one substance present at equilibrium if the concentrations of the other species are known. Such a computation is shown in Example 10.1.

Example 10.1

Program to compute the equilibrium concentration of SO_3 at $900^\circ C$. Necessary information: $K_{900^\circ C} = 4,129$.

```
100  REM PROGRAM TO COMPUTE [SO3] GIVEN [SO2] AND [O2]
103  REM K IS GIVEN AT 900 DEG C.
105  LET K=4129
110  PRINT "ENTER [SO2] IN MOLES/LITER";
120  INPUT S2
130  PRINT "ENTER [O2] IN MOLES/LITER";
140  INPUT O2
150  LET S3=SQR(K*S2↑2*O2)
155  PRINT
160  PRINT "[SO3]=";S3;"MOLES/LITER"
170  END
```

```
RUN
ENTER [SO2] IN MOLES/LITER?3
ENTER [O2]  IN MOLES/LITER?5
[SO3]= 431.Ø51    MOLES/LITER
DONE
```

Note that the concentrations of the compounds are in moles per liter, even though the materials in the example are gases. For gases, one could also write the equilibrium expression in terms of partial pressures. However, the value of the equilibrium constant when pressures are used will not be the same, in general, as when concentrations are employed.

A more interesting example arises if we consider what happens when SO_2 and O_2 are mixed together. Obviously, the equilibrium expression is not at first satisfied, because $[SO_3] = 0$. Thus, SO_2 and O_2 will react to form SO_3 until

$$\frac{[SO_3]^2}{[SO_2]^2[O_2]} = 4,129$$

The question we ask is this: "What are the final concentrations of SO_3, SO_2, and O_2 when the system comes to equilibrium?"

To approach this problem, let us call the starting concentration of SO_2 A; and let us call the starting concentration of O_2 B. When the system has reached equilibrium, let us say x moles per liter of O_2 will have been consumed in the reaction. Looking at the chemical equation for the reaction, we see that 2x moles/liter of SO_2 would also be consumed, and 2x moles/liter of SO_3 would be formed. Therefore, at equilibrium it would be true that

$$\frac{(2x)^2}{(A - 2x)^2(B - x)} = 4,129$$

or

$$\frac{(2x)^2}{(A - 2x)^2(B - x)} - 4,129 = 0 \qquad\qquad (10.1)$$

To find the equilibrium concentrations, we need to solve this equation for x; then

$$[SO_3] = 2x$$
$$[SO_2] = A - 2x$$
$$[O_2] = B - x$$

Solving this equation by hand exactly would be quite a task, for it is a third-degree (cubic) polynomial equation.

Problem 10.1

Show that Equation 10.1 is a third-degree equation. Would the computer be useful in proving this?

10.2 Solving the Polynomial Equation

The polynomial equations arising from equilibrium problems can sometimes be solved by hand by making certain simplifying approximations. While the techniques of approximation for hand solution are quite useful, it is nearly as easy to solve the equations without simplification when one has a computer available. Your computer may have in its program library a program or subroutine for finding the roots of polynomial equations.

There are a number of different computer techniques for solving polynomial equations. These methods can be divided into two major classes:

1 closed-form methods

2 iterative methods

In the closed-form approach, an analytic expression for the roots of the equation is incorporated into the computer program. For instance, we know that the roots of a second-degree (quadratic) equation $ax^2 + bx + c = 0$ are

$$x = \frac{-b \pm \sqrt{b^2 - 4ac}}{2a}$$

The program would take values for a, b, and c as input data and then evaluate the analytic expression. This approach is certainly simple from a programming point of view, but is practical only for quadratic, and perhaps cubic, equations because closed-form solutions for higher degree equations are not available in general.

It is clear that one pass through the program described in the previous paragraph would yield the correct roots for the quadratic equation. In an iterative solution, we first guess a root for the equation--sometimes arbitrarily--and then make repeated (hence, "iterative") attempts to refine the solution until no further

improvement can be achieved. There are a number of strategies for improving the
initial guess so that we converge upon a true solution of the equation we are interested in. These strategies include the method of successive approximations, the
Newton-Raphson method, and the secant method.* Each of these methods is capable
of solving general polynomial equations; each has particular advantages and weak-
nesses.

Rather than discuss the details of these various standard methods, we shall
employ here a very simple iterative procedure which, though not suited for general
use, will work for us because of the special constraints upon the roots of an equation in a chemical equilibrium problem. For example, the polynomial arising from
the SO_2 problem (Equation 10.1) is a cubic equation and has three roots. Only <u>one</u>
of these roots can be the answer to our problem, because the concentration of a
chemical cannot have three values at once but must have some particular value. We
immediately realize that negative, imaginary or complex roots cannot represent concentrations. We also know that x, the solution, could not possibly be larger than
the initial concentration of the O_2, because x represented the amount of O_2 that
reacted to achieve equilibrium. This means that, of all the possible values for
roots of our equation, the root we are looking for must lie between O and B. This
certainly narrows the field! Finally, as we have already said, there can be only
one root in this range if the equation is to be chemically meaningful. Thus, we
can reduce our task to finding the number between O and B that satisfies
Equation 10.1.

What do we mean when we ask if a particular value of x satisfies Equation 10.1?
The form of Equation 10.1 is

$$f(x) = 0$$

If a value of x (call it x_o) satisfies this equation, then $f(x)$ will equal zero
when x_o is substituted into the left-hand side of the equation. For values of x
that differ from x_o, $f(x)$ will be greater than or less than zero. This provides the
key to our strategy, for as one examines successive values of x near the root x_o,
the sign of $f(x)$ will switch from + to - (or vice versa) as we pass from $x < x_o$ to
$x > x_o$. Thus, we can locate x_o by looking for a change in sign in $f(x)$ as we

*See, for example, K. J. Johnson, "Numerical Methods in Chemistry," University of
 Pittsburgh Press, 1972.

systematically scan the range of possible values for x. This approach would not be
suitable for general use where x_o could be a positive or negative real, imaginary,
or complex number of any magnitude. But for our problem, we know the solution must
be a real number between 0 and B and that there can be only one solution in this
range. Thus, we shall examine the range $0 < x < B$ in increments, say, of 0.01 mole/
liter. This search will yield two values, x_1 and x_2, such that $f(x_1) < 0$ and
$f(x_2) > 0$ (or vice versa). x_1 and x_2 will differ by 0.01 mole/liter, and we will
know that x_o lies between x_1 and x_2. Then, to refine our estimate of x_o, we shall
subdivide the range x_1 to x_2 into 100 parts and search this range to find out more
precisely where the sign change occurs. This search will yield two values, x_3 and
x_4, such that $f(x_3) < 0$ and $f(x_4) > 0$ (or vice versa). x_3 and x_4 will differ by
0.0001, and we will know that x_o lies between x_3 and x_4. By performing this procedure iteratively, we can close in on the value of x_o to any desired degree of precision within the inherent precision of the computer being used. This approach is
shown pictorially in Figure 10.1.

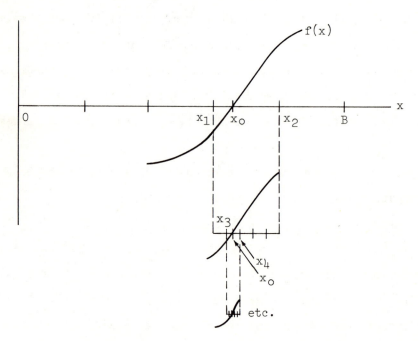

Figure 10.1   Pictorial representation of subdivision method for finding a root.

Example 10.2 contains a flow chart and program incorporating the subdivision method
for finding the root.

Example 10.2

Program to solve polynomial equations arising from chemical equilibrium problems.

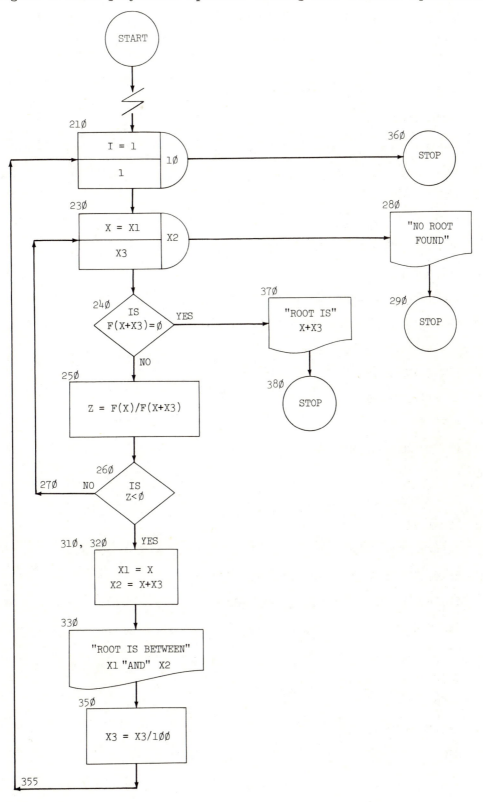

Figure 10.2

```
100   REM PROGRAM TO SOLVE POLYNOMIAL EQUATIONS FROM EQUILIBRIUM PROBLEMS
110   REM FUNCTION DEFINITION FOR SO2 PROBLEM
120   DEF FNA(X)=(2*X)↑2/((A-2*X)↑2*(B-X))-4129
130   PRINT "ENTER INITIAL CONCENTRATION OF SO2";
140   INPUT A
150   PRINT "ENTER INITIAL CONCENTRATION OF  O2";
160   INPUT B
165   PRINT
170   LET X1=0
180   LET X2=50
190   LET X3=.01
200   REM RUN FOR TEN CYCLES IF NECESSARY
210   FOR I=1 TO 10
220   REM SEARCH INTERVAL FOR SIGN CHANGE
230   FOR X=X1 TO X2 STEP X3
240   IF FNA(X+X3)=0 THEN 370
250   LET Z=FNA(X)/FNA(X+X3)
260   IF Z<0 THEN 310
270   NEXT X
280   PRINT "NO ROOT FOUND"
290   STOP
300   REM SPECIFY NEW INTERVAL
310   LET X1=X
320   LET X2=X+X3
330   PRINT "ROOT IS BETWEEN ";X1;"AND ";X2
335   IF ABS((X2-X1)/X2)<.00001 THEN 380
340   REM SPECIFY NEW INCREMENT SIZE
350   LET X3=X3/100
355   NEXT I
360   STOP
370   PRINT "ROOT IS ";X+X3
380   END
```

The fundamental search for a sign change in f(x) is carried out in lines 230 to 270. Line 240 takes care of the possibility that we might land exactly on the root in one of our passes. Lines 310 to 330 establish the bounds of the subinterval to be searched after a sign change has been found; and line 350 establishes the reduced increment to be used in searching the new subinterval. This procedure tends to converge on the root in rather few iterations. X1 becomes very close to X2; because of small errors in decimal number representation in the computer, the FOR/NEXT loop on X can become an infinite loop. To avoid this, we add a line which terminates the computation when X1 and X2 become so close in value that the computer cannot accurately distinguish them. The actual tolerance here will depend on the computer you are using. For a machine that displays six significant figures, 1×10^{-5} seems to be an appropriate value. Thus, we add one line to the above program:

335 IF ABS((X2-X1)/X2) < 1.E-∅5   THEN 38∅

With X2 set equal to 50, the program cannot find a root larger than 50. Depending on the problem you are solving, it might be necessary to change the value of X2 to include the solution. In some cases, it may be desirable to increase the value of X3 to speed the running of the program, but if X3 is too large, the program may miss the solution altogether.

```
RUN
ENTER INITIAL CONCENTRATION OF S∅2?3.∅∅∅
ENTER INITIAL CONCENTRATION OF ∅2?5.∅∅∅
ROOT IS BETWEEN  1.48        AND  1.49
ROOT IS BETWEEN  1.48759     AND  1.48769
ROOT IS BETWEEN  1.48765     AND  1.48765
DONE
```

In this sample RUN, we have set the initial concentration of SO_2 equal to 3 mole/liter and the initial concentration of O_2 equal to 5 mole/liter. The program solves Equation 10.1 and tells us that x = 1.488. Finally, recalling our definition for x, we find that at equilibrium

$$[SO_3] =   2x = 2.976 \text{ mole/liter}$$
$$[SO_2] = A - 2x = (3.000 - 2.976)\text{mole/liter} = 0.024 \text{ mole/liter}$$
$$[O_2] = B - x = (5.000 - 1.488)\text{mole/liter} = 3.512 \text{ mole/liter}$$

How do we know whether these values are correct? Because it is unwise to put total trust in a computer or in a computer program, let us check these concentrations by substituting them back into the equilibrium expression to see if they in fact satisfy that expression.

$$\frac{[SO_3]^2}{[SO_2]^2[O_2]} = \frac{2.976^2}{0.024^2 \times 3.512} = 4,378$$

This result agrees poorly with the actual equilibrium constant of 4,129. Can it be that the computer gives such inaccurate answers?

Closer examination shows that the computer is not at fault. Notice that the least precise datum in the above check, 0.024, has only two significant figures. Let us redo the check using the more precise numbers given in the computer output. (Bear in mind that these values are solutions to an equation, not laboratory measurements, so we are justified in taking all six figures as significant.)

$$\frac{2.97530^2}{0.02470^2 \times 3.51235} = 4,131$$

Now the agreement with K = 4,129 is far better. Take note that the weakest spot is still the 0.02470, with four significant figures. This datum results from the subtraction of two numbers which, though precise to six places themselves, are quite close together in value. The subtraction then leads to a loss of precision. This sort of situation is often unavoidable, and one must be aware of its existence and of the effect on the overall precision of a computation.

Problem 10.2

Use the program in Example 10.2 to find the concentrations of SO_2, O_2, and SO_3 for the following initial concentrations of SO_2 and O_2:

	$[SO_2]$ initial	$[O_2]$ initial
(a)	1 mole/liter	1 mole/liter
(b)	2 mole/liter	2 mole/liter
(c)	1 mole/liter	10 mole/liter
(d)	10 mole/liter	1 mole/liter
(e)	10 mole/liter	10 mole/liter

Problem 10.3

Modify the program of Example 10.2 to solve the following equilibrium:

$$N_2 + 3H_2 \rightleftharpoons 2NH_3$$

Necessary information: $K = 0.1510$ at $450^\circ C$.

Compute the concentrations of the three species given the following initial concentrations:

	$[N_2]$ initial	$[H_2]$ initial
(a)	1 mole/liter	1 mole/liter
(b)	2 mole/liter	2 mole/liter
(c)	1 mole/liter	10 mole/liter
(d)	10 mole/liter	1 mole/liter
(e)	10 mole/liter	10 mole/liter

10.3 Equilibria Involving Solids

An interesting situation arises when one of the materials involved in an equilibrium is a solid. The "concentration" of a solid (or of a pure liquid) is determined solely by its density. Thus, as long as the solid is present, its concentration does not change. For this reason, solid concentrations do not appear in the equilibrium expression; instead, the constant value is combined into the equilibrium constant. For example, for the reaction

$$CaCO_3 \rightleftharpoons CaO + CO_2$$

the equilibrium expression is just

$$[CO_2] = K$$

because both $CaCO_3$ and CaO are solids.

A special instance of equilibria involving solids is the dissolving of a solid in a solvent. For example,

$$CaSO_4 \ (+ \text{ lots of water}) \rightleftharpoons Ca^{+2}_{aq} + SO^{-2}_{4\ aq} \ (+ \text{ lots of water})$$

Here, the essential equilibrium is

$$CaSO_4 \rightleftharpoons Ca^{+2}_{aq} + SO^{-2}_{4\ aq}$$

Of course, water is necessary for this process to take place, but the water is present in large excess, and its concentration remains effectively constant. Thus, the equilibrium expression is

$$[Ca^{+2}_{aq}][SO^{-2}_{4\ aq}] = K_{sp}$$

(Why does $[CaSO_4]$ not appear in this expression?)

In this kind of equilibrium expression, the constant is called the "solubility product". There are several immediate generalizations we can make about the solubility product:

1 If an ionic solid is in equilibrium with its dissolved ions, the product of the ion concentrations (raised to the appropriate powers) will be equal to K_{sp} for that material.

2 If a solution exists in which the product of the ion concentrations (raised to the appropriate powers) equals K_{sp} for some material composed of those ions, then the solid form of the material must also be present and an additional amount of any of the ions will cause additional solid to precipitate.

3 If a solution exists in which the product of the ion concentrations (raised to the appropriate powers) is less than K_{sp} for some material composed of those ions, then that material will not be present as a solid.

This information is used in the following example.

Example 10.3

Program to determine whether a substance will precipitate from water solution. Necessary information: solubility product of the substance.

```
1ØØ   REM PROGRAM TO DETERMINE WHETHER A SUBSTANCE WILL PRECIPITATE
1Ø5   LET P=1
11Ø   PRINT "ENTER SOLUBILITY PRODUCT";
12Ø   INPUT K
13Ø   PRINT "ENTER FOR EACH ION: MOLARITY OF ION, FORMULA SUBSCRIPT"
134   PRINT "   (END FLAG IS 999,999)"
136   PRINT
14Ø   PRINT "CONC., SUBS."
145   PRINT
15Ø   INPUT C,S
16Ø   IF C=999 THEN 188
17Ø   LET P=P*C↑S
18Ø   GOTO 15Ø
188   PRINT
```

```
19Ø   IF P >= K THEN 22Ø

2ØØ   PRINT "SOLID WILL NOT PRECIPITATE"

21Ø   GOTO 3ØØ

22Ø   PRINT "SOLID WILL PRECIPITATE"

3ØØ   END

RUN

ENTER SOLUBILITY PRODUCT?1.E-1Ø

ENTER FOR EACH ION: MOLARITY OF ION, FORMULA SUBSCRIPT

     (END FLAG IS 999,999)

CONC., SUBS.

?.ØØ1,1

?.ØØ2,1

?999,999

SOLID WILL PRECIPITATE

DONE
```

This program is quite simple. Line 150 INPUTs the concentration of each ion as well as the power to which that concentration must be raised. The product of the concentrations, raised to the appropriate powers, is accumulated in line 170. Note the use of the "flag" to break the INPUT loop.

Line 190 determines whether the product of the concentrations (raised to the appropriate powers) exceeds the solubility product for the material in question and, thus, whether the material will precipitate.

Problem 10.4

Listed below are the solubility products for a number of common sulfide compounds. Write a program that will

1 ask the user to specify the compound he is interested in and the concentrations of metal ion and sulfide ion present;

2 determine whether or not the solid compound will precipitate out of solution.*

*Caution: some computers may not be able to handle a number as small as 10^{-50}; such machines will generate an UNDERFLOW error if you try to run such examples. To get around this limitation, take the square root of the whole equilibrium equation and work with the resulting formula instead of the original equilibrium equation.

Compound	K_{sp}, $25^{\circ}C$
Ag_2S	7×10^{-50}
CdS	7×10^{-27}
CuS	8×10^{-36}
FeS	5×10^{-18}
HgS	3×10^{-52}
PbS	8×10^{-28}

Problem 10.5

Modify the program from 10.4 so that it

1 asks the user to specify the compound and the concentration of metal ion;

2 determines the concentration of sulfide ion that must be present to precipitate the solid metal sulfide.

Problem 10.6

Figure out how to calculate the solubility of a substance in moles per liter, given the K_{sp} and the chemical formula. Is this task easier to do by hand or with the computer? Under what circumstances might you write a computer program for this job?

Problem 10.7

Use the method you developed in Problem 10.6 to calculate the solubility in moles per liter of

Compound	K_{sp}, $25^{\circ}C$
$AgCl$	1×10^{-10}
CaF_2	4×10^{-11}
FeS	5×10^{-18}

10.4 Acid-Base Equilibria

A ubiquitous and important class of chemical equilibria involves solutions containing acids and bases. Let us first clarify some terms.

Acid. There are various chemical definitions of "acid." Depending on the situation, one or another of these definitions may be appropriate. For the problems we are about to study, we shall define an acid as a chemical species that can give up H^+.

Base. There are also various definitions of "base," corresponding to the definitions of "acid." In keeping with the foregoing definition of acids, let us define
bases as those chemical species which can pick up H^+.

These are called the "Brønsted" definitions of acids and bases and are the
definitions in most common use.

It is important to realize that these definitions are not mutually exclusive.
The species HCO_3^-, for example, can act as an acid by giving up H^+:

$$HCO_3^- \rightarrow H^+ + CO_3^{-2}$$

The same ion can also function as a base by picking up H^+:

$$HCO_3^- + H^+ \rightarrow H_2CO_3$$

If an acid loses a proton (H^+)

$$HA \rightarrow H^+ + A^-$$

then, clearly, the resulting species (A^-) must be capable of gaining the proton back

$$A^- + H^+ \rightarrow HA$$

Thus, A^- satisfies our definition of a base; A^- is referred to as the "conjugate
base" of the acid HA. The pair of reactions comprises an equilibrium system.

Similarly, if a base B picks up a proton

$$B + H^+ \rightarrow BH^+$$

then the resulting species (BH^+) must be considered the "conjugate acid" of the
base B.

The "strength" of an acid is the extent to which the acid tends to foist its
protons upon other molecules. The strength of a base is the extent to which the
base strips protons from other molecules. Acid and base strengths are measured by
the equilibrium constant for the process in question:

(1) $HA \rightleftharpoons H^+ + A^-$

$$\frac{[H^+][A^-]}{[HA]} = K_A \quad \text{(sometimes called the dissociation constant)}$$

(2) $B + H^+ \rightleftharpoons BH^+$

$$\frac{[BH^+]}{[B][H^+]} = K'_B$$

There is a subtle catch to these two expressions. Process 1 can occur only if
there is some species present to accept the proton from HA. Process 2 can

occur only if there is some species that can donate a proton to B.   Suppose,
then, that Z is a species that can accept a proton.   Then, if HA is mixed
with Z, we would have

(3) $HA + Z \rightleftharpoons HZ^+ + A^-$

$$K = \frac{[HZ^+][A^-]}{[HA][Z]}$$

Notice that

$$K = \frac{[HZ^+][A^-]}{[HA][Z]} = \left(\frac{[H^+][A^-]}{[HA]}\right)\left(\frac{[HZ^+]}{[H^+][Z]}\right)$$

which is just the product of the equilibrium constant for HA acting as an
acid (K_A) and the equilibrium constant for Z acting as a base.   Therefore,
the overall equilibrium constant for the acid-base reaction (Process 3)
depends on both the strength of HA as an acid and the strength of Z as a base.

10.5 Acid-Base Behavior of Water

According to our definitions of acids and bases, water can be both an acid
and a base.   The dissociation of water to yield H^+ and OH^-

$$H_2O \rightleftharpoons H^+ + OH^-$$

$$K_W = [H^+][OH^-] = 1 \times 10^{-14}$$

occurs only to a very small extent, as indicated by the small value of the equilibrium constant K_W.

Although water is not a strong base, any free H^+ produced by an acid dissolved
in water will be taken up by the water as follows:

$$H_2O + H^+ \rightleftharpoons H_3O^+$$

Thus, H^+ in aqueous solution is frequently written as H_3O^+.

When a _base_ is dissolved in water, the proton to be acquired by the base comes
from the water.

(4) $B + H_2O \rightleftharpoons BH^+ + OH^-$

$$K_B = \frac{[BH^+][OH^-]}{[B]}$$

This is the K_B that is found in reference tables.   It is related in a simple
way to K_B' from Process 2:

$$K_B = \frac{[BH^+][OH^-]}{[B]} = \frac{[BH^+][OH^-][H^+]}{[B][H^+]} = K_B' K_w$$

10.6   pH

The concentration of H^+ in aqueous solution is often expressed in pH units. The pH of a solution is defined as:

$$pH = -\log[H^+]$$

Thus, if a solution has an H^+ concentration of 2.000×10^{-5} mole/liter, then the pH will be

$$pH = -\log(2 \times 10^{-5}) = -(-4.699) = 4.699$$

The pH scale provides a convenient way of condensing a very wide range of concentrations (say, 0.1 M to 10^{-14} M) into a more tractable numerical range (1 to 14).

10.7   Calculations of Acid-Base Equilibria

Knowledge of the equilibrium state of a chemical system is important because all systems tend naturally to seek equilibrium. (In some cases, though, the approach to equilibrium may be so slow as to be imperceptible.) Thus, we often desire to calculate ahead of time the final equilibrium concentrations that will exist in a chemical system after we mix the various components together. For example, if a nutrient medium for bacteria were being prepared that required the addition of Na_2HPO_4, we might want to know what the actual concentrations of PO_4^{-3} and H^+ would be in the resulting solution. There are various types of chemical equilibria involving solubility, complexation, oxidation-reduction, and acid-base reactions. We shall deal here with the computational details of acid-base systems. Similar considerations apply to other equilibria.

As we have already seen, conceptually simple equilibrium problems frequently lead to equations that are difficult to solve by hand. Your textbook probably discusses techniques for solving such equations through the use of judicious approximations. The ability to do that sort of quick calculation is a valuable asset and aids one's chemical intuition.* However, we must also realize that the digital

---

*An excellent treatment of such methods will be found in James N. Butler, "Solubility and pH Calculations," Addison-Wesley, 1964.

computer makes it possible to solve these equations without any approximations and
with hardly more effort than that required for the approximate approach.

The first step in the exact treatment of an equilibrium problem is to write
down all the relevant chemical relationships. These relationships fall into three
classes:

    1   equilibrium equation(s)

    2   mass balance equation(s)

    3   charge balance equation

The number of equations you write down must be equal to the number of unknown concentrations involved in the problem.

Let's plunge right into the deep end. Here is a problem that would cause a
nonprogramming student to blanch: calculate exactly the hydrogen ion concentration
in a 0.05 M solution of oxalic acid. Oxalic acid, $H_2C_2O_4$, has two hydrogen atoms
that can dissociate in keeping with our definition of an acid. However, both
hydrogen atoms do not have the same tendency to dissociate. Let us call the
undissociated acid H_2A. Then

$$H_2A \rightleftharpoons H^+ + HA^- \qquad \frac{[H^+][HA^-]}{[H_2A]} = K_1 = 6.3 \times 10^{-2} \qquad (10.2)$$

$$HA^- \rightleftharpoons H^+ + A^{-2} \qquad \frac{[H^+][A^{-2}]}{[HA^-]} = K_2 = 6.1 \times 10^{-4} \qquad (10.3)$$

Because we are working in a water solution, the water dissociation equilibrium must
hold

$$H_2O \rightleftharpoons H^+ + OH^- \qquad [H^+][OH^-] = 1 \times 10^{-14} \qquad (10.4)$$

These three equations are the equilibrium equations which govern this particular
system. We must now write the mass balance and charge balance equations.

The mass balance equation acknowledges the fact that, if we dilute C moles of
H_2A with water to make one liter of solution, then no matter how the H_2A dissociates, associates, or whatever, the total concentration of the species containing A
(in this case, oxalate) must add up to C, the original amount. Thus,

$$[H_2A] + [HA^-] + [A^{-2}] = C \qquad (10.5)$$

The charge balance equation embodies the fact that the total charge due to
positive ions in solution must equal the total negative charge. In the present
example, the only positive species is H^+, and the negative ions are HA^-, A^{-2}, and

OH^-. Then

$$[H^+] = [HA^-] + 2[A^{-2}] + [OH^-] \tag{10.6}$$

These five equations are the ones needed to solve the problem. We now need to take these five equations and, by substitution, eliminate all unknowns except $[H^+]$. If this task is approached systematically, it is not as terrifying as it might seem. Consider the following five steps:

(1) First, consider only the equilibrium equations and the mass balance equation. Choose an unknown (not $[OH^-]$) that appears least frequently in these equations.* Solve for that unknown in the equilibrium equation in which it appears and substitute the resulting expression into the mass balance equation. In our oxalic acid case, we could choose either $[H_2A]$ or $[A^{-2}]$. If we choose $[H_2A]$, then from Equation 10.2 we would have

$$[H_2A] = \frac{[H^+][HA^-]}{K_1}$$

We substitute this into the mass balance Equation 10.5.

$$\frac{[H^+][HA^-]}{K_1} + [HA^-] + [A^{-2}] = C$$

(2) We repeat Step 1 until the mass balance equation contains only $[H^+]$ and one other unknown.

In our example, the next species appearing infrequently is A^{-2}. We solve for $[A^{-2}]$ using the equilibrium Equation 10.3,

$$[A^{-2}] = \frac{K_2[HA^-]}{[H^+]}$$

and substitute this into the mass balance equation. The mass balance equation now looks like this, containing only $[H^+]$ and $[HA^-]$:

$$\frac{[H^+][HA^-]}{K_1} + [HA^-] + \frac{K_2[HA^-]}{[H^+]} = C$$

(3) Take the modified mass balance from Step 2 and solve for the unknown other than $[H^+]$. This can be done by factoring. In our present example, we would have

---

*This strategy is owed in part to James N. Butler's work, op. cit.

$$[HA^-] = \cfrac{C}{\cfrac{[H^+]}{K_1} + 1 + \cfrac{K_2}{[H^+]}}$$

(4) Now take up the charge balance Equation 10.6. Make the same substitutions into the charge balance equation that you made into the mass balance equation in Steps 1 and 2, until the charge balance equation contains the same two unknowns as the modified mass balance equation.

In the present case, this would give

$$[H^+] = [HA^-] + 2\,\frac{K_2[HA^-]}{H^+} + [OH^-]$$

$$= [HA^-] + 2\,\frac{K_2[HA^-]}{[H^+]} + \frac{K_w}{[H^+]}$$

(Note that $[OH^-]$ can always be eliminated by using the equilibrium equation for water.)

(5) At this point, you will have two equations in two unknowns; one of these unknowns is $[H^+]$. To obtain finally an equation containing only $[H^+]$, substitute the equation from Step 3 into the equation from Step 4.

For the oxalic acid example we have been following, this will give

$$[H^+] = \cfrac{C}{\cfrac{[H^+]}{K_1} + 1 + \cfrac{K_2}{[H^+]}} + \cfrac{2\,K_2 C}{\cfrac{[H^+]^2}{K_1} + [H^+] + K_2} + \cfrac{K_w}{[H^+]} \qquad (10.7)$$

To find the hydrogen ion concentration in our 0.05 M oxalic acid solution, we let $C = 0.05$, $K_1 = 6.3 \times 10^{-2}$, $K_2 = 6.1 \times 10^{-4}$, $K_w = 1 \times 10^{-14}$, and we solve the equation for H^+.

The prospect of solving this complicated polynomial equation need not cause panic, for we have already developed the apparatus for solving such equations. If we rewrite Equation 10.7 as follows,

$$[H^+] - \cfrac{C}{\cfrac{[H^+]}{K_1} + 1 + \cfrac{K_2}{[H^+]}} - \cfrac{2\,K_2 C}{\cfrac{[H^+]^2}{K_1} + [H^+] + K_2} - \cfrac{K_w}{[H^+]} = 0$$

the whole equation takes on the form

$$f([H^+]) = 0$$

To solve this equation, we can use the strategy and the major part of the program from Example 10.2. A minor modification does need to be made in the program. Our present polynomial contains the unknown in the denominator of a fraction in certain places. If our program allows the search for the root to begin at x = 0 (line 170 LET X1 = $\emptyset$), then we will generate DIVIDE-BY-ZERO errors in the computer. To avoid this problem, we can make X1 equal to some very small but nonzero value, such as X1 = 1.E - 20. This change, however, engenders another danger. Every computer has an upper and lower limit on the size of the numbers it can represent. Suppose the range for a particular computer is 1.E + 38 to 1.E - 38. When that computer tries to generate the X↑2 necessary to evaluate our function, the result would be 1.E - 40 if X1 = 1.E - 20, and we would generate an UNDERFLOW error. Therefore, you will have to be careful to use a value of X1 such that you do not underflow the computer you are working with. By making X1 a number other than zero, there is the danger that we will miss the root entirely; the program itself takes into account this eventuality (line 280). That the root is outside the numerical range accessible to the computer is a remote possibility, but it underscores the fact that computers are, after all, machines with finite limitations.

Problem 10.8

Modify the program from Example 10.2 to solve the oxalic acid equilibrium for a 0.05 M solution. Find the pH of this solution.

Problem 10.9

Use your program from the previous problem to find the pH of a 0.1 M solution of oxalic acid. Repeat for a 0.2 M solution.

Problem 10.10

Acetic acid $(HC_2H_3O_2)$ is a monoprotic acid. The dissociation constant is K = 1.85 x 10^{-5}. Derive the equation for $[H^+]$ using Steps 1 through 5 and modify your program from the previous problem to solve the acetic acid equilibrium. Use your program to compute the pH of the following acetic acid solutions:

(a) 0.000 M

(b) 0.001 M

(c) 0.005 M

(d) 0.010 M

(e) 0.100 M

(f) 1.000 M

Problem 10.11

Phosphoric acid (H_3PO_4) is a triprotic acid. The dissociation constants are

$K_1 = 5.89 \times 10^{-3}$ $K_2 = 6.16 \times 10^{-8}$ $K_3 = 4.79 \times 10^{-13}$. Derive the equation for

$[H^+]$ using Steps 1 through 5. Modify your program to solve the phosphoric acid

equilibrium for the following solutions:

(a) 0.001 M

(b) 0.010 M

(c) 0.100 M

(d) 1.000 M

10.8 Hydrolysis of Salts

We have seen that when a weak acid HA is dissolved in water, it partially

dissociates to produce H^+ (H_3O^+):

$$HA + H_2O \rightleftharpoons H_3O^+ + A^-$$

Here, the water is acting as a base by taking a proton from HA. Water can also

behave as an acid, and when the ion A^- (the conjugate base of the weak acid HA) is

dissolved in water, the water will donate a proton, leaving OH^-:

$$A^- + H_2O \rightleftharpoons HA + OH^-$$

This means that, if we dissolve the salt of a weak acid, say NaA, in water, we will

get a basic solution. This reaction of the salt with water is called hydrolysis.

As an example, we will choose the salt of oxalic acid, our weak acid example.

You might think that we are about to get into another complicated analysis

like the one we went through to get the $[H^+]$ equation for oxalic acid. A little

thought, however, shows that the new situation is very simply related to the previ-

ous one. Because the only new species in solution is Na^+, which is completely dissociated in solution, the <u>equilibrium equations</u> governing the Na_2A system are exactly the same as the ones that applied to H_2A (Equations 10.2, 10.3, 10.4). Also, because the metal ion separates completely from the oxalate ion, the mass balance equation for oxalate will be the same as before (Equation 10.5).

In fact, the <u>only</u> difference introduced by using Na_2A instead of H_2A is in the charge balance equation, for now we have an additional positive species in solution (Na^+). The charge balance relation therefore becomes

$$[Na^+] + [H^+] = 2[A^{-2}] + [HA^-] + [OH^-]$$

or

$$2C + [H^+] = 2[A^{-2}] + [HA^-] + [OH^-]$$

Looking back at our derivation of the H_2A equation for $[H^+]$ (Steps 1 through 5), we will see that Steps 1 through 3 will go exactly the same as before. In Step 4, the charge balance equation will have 2C added to the left hand side, leading to

$$2C + [H^+] = \cfrac{C}{\cfrac{[H^+]}{K_1} + 1 + \cfrac{K_2}{[H^+]}} + \cfrac{2 K_2 C}{\cfrac{[H^+]^2}{K_1} + [H^+] + K_2} + \cfrac{K_w}{[H^+]} \qquad (10.8)$$

in Step 5. This equation is identical with Equation 10.7 except for the 2C added to the left-hand side.

Thus, we can use the same program to calculate $[H^+]$ for solutions of the salts of weak acids simply by making a minor change in the DEF statement, which establishes the form of the equation being solved.*

Problem 10.12

Modify your program from Problem 10.8 to treat solutions of $Na_2C_2O_4$. Find the pH of the following solutions of $Na_2C_2O_4$:

*NOTE: We have assumed that the cation does not undergo hydrolysis. If this assumption were false, as would be the case if the salt were formed from a weak acid and a weak base, e.g., $(NH_4)_2C_2O_4$, then we would have to incorporate the appropriate new equilibrium equations for the cation, add a mass balance equation for the base, and include the new species in the charge balance equation. The procedure outlined in Steps 1 through 5 would then lead to an equation for $[H^+]$ which could be solved with the program already described.

(a)   0.05 M

(b)   0.10 M

(c)   0.20 M

Problem 10.13

Calcium oxalate is sparingly soluble in water; $K_{sp} = 2 \times 10^{-9}$. Calculate the pH of a saturated solution of this salt. Assume that calcium hydroxide is completely dissociated.

Problem 10.14

Modify your program from Problem 10.10 to treat solutions of $NaC_2H_3O_2$. Compute the pH of sodium acetate solutions of the same concentrations as given in that problem.

Problem 10.15

Modify your program from Problem 10.11 to treat solutions of

(a) Na_3PO_4

(b) Na_2HPO_4

(c) NaH_2PO_4

Compute the pH of solutions of these salts using the same series of concentrations as those given in that problem.

10.9   Titration Curves

Titration is the process of adding measured amounts of one solution to another solution. This process is often used in chemical analysis to find the concentration of some species, by titrating with a reactant solution of known concentration.

In titrations, it is sometimes useful to record the whole course of the titration rather than simply to record the end point. Such a titration record is called a titration curve; a titration curve is a plot of the concentration of some chemical species (such as H^+ in the case of acid-base titrations) versus volume of titrant added. Such a plot gives an overall picture of the titration and gives a graphic view of the inherent sharpness of the end point.

It is useful to be able to anticipate the general shape of titration curves for various kinds of situations. We might want to calculate what the titration curve would look like for various concentrations of a particular acid and base. We can do this handily using the techniques developed in the last two sections.

Suppose we have a beaker containing 50 ml of a 0.1M solution of $H_2C_2O_4$. We already know how to find the equation for $[H^+]$ for this situation and how to solve this equation. We have also seen that, if we switch from $H_2C_2O_4$ to $Na_2C_2O_4$, the only change is in the charge balance equation, which leads to a very simple change in the final equation. Then if we take our beaker of $H_2C_2O_4$ solution and begin to add portions of a solution of 0.1 M NaOH, the equilibrium equations and mass balance equation will still remain unchanged. The only effect on the equations governing the system will be the addition of $[Na^+]$ to the charge balance equation and the decrease in C (the formal concentration of the oxalic acid) due to the increase in total volume of liquid in the beaker which accompanies the addition of the NaOH solution. With these modifications, the equation for $[H^+]$ becomes

$$[Na^+] + [H^+] = \frac{C}{\frac{[H^+]}{K_1} + 1 + \frac{K_2}{[H^+]}} + \frac{2 K_2 C}{\frac{[H^+]^2}{K_1} + [H^+] + K_2} + \frac{K_w}{[H^+]} \qquad (10.9)$$

where C is no longer constant but changes with each addition of NaOH solution. To produce a titration curve, we must re-solve this equation for each addition of titrant. If we let

V_1 = initial volume of $H_2C_2O_4$ solution,

C_1 = initial formal concentration of $H_2C_2O_4$,

V_2 = cumulative volume of titrant added,

C_2 = concentration of titrant,

then, in our present example,

$$[Na^+] = \frac{\text{moles of NaOH added}}{\text{total volume in beaker}} = \frac{C_2 V_2}{V_1 + V_2}$$

and

$$C = \frac{\text{moles of } H_2C_2O_4 \text{ originally present}}{\text{total volume in beaker}} = \frac{C_1 V_1}{V_1 + V_2}$$

Putting this all together, the equation for $[H^+]$ at any point in the titration becomes

$$\frac{C_2 V_2}{V_1 + V_2} + [H^+] = \frac{\dfrac{C_1 V_1}{V_1 + V_2}}{\dfrac{[H^+]}{K_1} + 1 + \dfrac{K_2}{[H^+]}} + \frac{2K_2 \dfrac{C_1 V_1}{V_1 + V_2}}{\dfrac{[H^+]^2}{K_1} + [H^+] + K_2} + \frac{K_w}{[H^+]} \qquad (10.10)$$

For the example we have been discussing, we would have

C_1 = 0.1 mole/liter,

V_1 = 0.05 liter,

C_2 = 0.1 mole/liter.

For each addition of NaOH solution, we would substitute the appropriate value for V_2 and solve the equation for H^+ using the program we have already devised.

Problem 10.16

Modify your program from Problem 10.12 to compute points on a titration curve.

(a) Use the modified program to compute the points on a titration curve for the titration of 50ml of 0.1 M $H_2C_2O_4$ with 0.1 M NaOH. It is suggested that you compute points in increments of, say, 2 ml and plot (pH) versus (ml added) on a piece of graph paper. Then go back to the computer and compute points to fill in the plot wherever you do not have enough points to draw a good curve.

(b) Do the same for 50ml of 0.1 M acetic acid titrated with 0.1 M NaOH.

(c) Do the same for 50ml of 0.2 M acetic acid titrated with 0.1 M NaOH.

(d) Do the same for 50 ml of 0.001 M acetic acid titrated with 0.001 M NaOH.

Compare these titration curves.

Problem 10.17

You can use the equation for diprotic acids in monoprotic acid problems by making K_2 = 0 and making K_1 equal to the monoprotic acid dissociation constant. You can use the same equation for strong acids, such as HCl and HNO_3 by making K_2 = 0 and K_1 quite large, say 1.E+04. (Show why this should be so.)

Use your program from Problem 10.16 to compute the titration curve for 50 ml of 0.1 M HCl titrated with 0.1 M NaOH. Compare this titration curve with those obtained in Problem 10.16. What is the pH at the endpoint in each case?

10.10 Buffers

A buffer is a solution that contains both a weak acid and a salt of the weak acid; or, a buffer solution can be made from a weak base and a salt of the weak base. The name "buffer" is used because such solutions resist changes in pH; that is, the H^+ concentration in a buffer solution changes relatively little when portions of strong acid or strong base are added.

We can examine this characteristic of buffers by studying the titration curve for a buffer solution titrated with a strong acid or base. In either case, only the charge balance and the value of C are affected. (See Equations 10.9 and 10.10.) Suppose we have a solution containing 0.1 mole/liter $H_2C_2O_4$ and 0.1 mole/liter $Na_2C_2O_4$. Looking back over the equations governing this system (pp. 179, 180, 183), we see that C in the mass balance equation will become $[H_2C_2O_4]_{formal}$ + $[Na_2C_2O_4]_{formal}$ = 0.2 mole/liter. In addition, the charge balance equation will have the term $[Na^+]$ on the left side, where $[Na^+]$ = $2x[Na_2C_2O_4]_{formal}$ = 0.2 mole/liter before any titrant has been added. If we take 50 ml of this solution and begin to titrate with 0.1 M NaOH, Equation 10.10 will be applicable with two modifications:

1 C_1 is now 0.2 mole/liter;
2 the first term on the left (representing $[Na^+]$) has to become

$$\frac{0.2 \text{ mole/liter} * V_1 + C_2 V_2}{V_1 + V_2}$$

to take into account the Na^+ present before the titration is begun.

Problem 10.18

Compare the titration curves for:

(a) 0.1 M $H_2C_2O_4$ titrated with 0.1 M NaOH

(b) 0.2 M $H_2C_2O_4$ titrated with 0.1 M NaOH

(c) a buffer solution made of 0.1 M $H_2C_2O_4$ and 0.1 M $Na_2C_2O_4$ titrated with 0.1 M NaOH.

11

Radiochemistry

In most chemical processes, we are concerned with changes in the structure of whole molecules, changes in the electron distribution within atoms, or changes in the spatial distribution of molecules (changes of state). In such processes, the atomic nuclei are unaltered.

11.1 Radiochemical Processes

In radiochemical processes, the nuclei themselves undergo various kinds of changes. A substance whose atomic nuclei decompose with consequent emission of particles or electromagnetic energy is said to be "radioactive". Such nuclear decompositions most commonly involve emission of an α particle (helium nucleus), a β^- particle (electron), or a γ-ray (an x-ray-like emission). In many instances, a different element is left as the result of the nuclear change.

Regardless of emission type, the <u>rate</u> of emission is proportional to the number of undecomposed nuclei present in a sample. Thus,

$$- \frac{dN}{dt} = kN$$

where $- \frac{dN}{dt}$ is the rate of disappearance of atoms of the original radioactive material, N is the number of atoms of the original radioactive material present at any given time, and k is a constant characteristic of the particular radioactive material.

188

Then,

$$\frac{dN}{N} = -k\ dt$$

and, integrating between N_o at $t = 0$ and N at t,

$$\int_{N_o}^{N} \frac{dN}{N} = -k \int_{o}^{t} dt$$

$$\ln \frac{N}{N_o} = -kt$$

$$\frac{N}{N_o} = e^{-kt}$$

The radioactive decay rate for a particular material is usually characterized by specifying the "half-life"; this is the amount of time required for 1/2 of the material originally present to decay. We can easily determine the relation between the half-life $\left(t_{1/2}\right)$ and k, since at $t = t_{1/2}$, $N = 1/2\ N_o$. Then

$$\frac{1/2\ N_o}{N_o} = e^{-kt_{1/2}}$$

$$1/2 = e^{-kt_{1/2}}$$

$$\ln 1/2 = -kt_{1/2}$$

$$t_{1/2} = -\frac{\ln 1/2}{k} = \frac{\ln 2}{k}$$

This relationship is the basis for Example 11.1.

Example 11.1

Program to compute the decay constant k for any radioactive element, given the half life.

```
8∅    REM PROGRAM TO COMPUTE  K  FROM HALF-LIFE
9∅    DIM A$[1∅]
1∅∅   PRINT "ENTER HALF LIFE";
11∅   INPUT T
12∅   PRINT "ENTER TIME UNITS - SECONDS, MINUTES, ETC.";
13∅   INPUT A$
14∅   LET K=LOG(2)/T
145   PRINT
```

```
15Ø   PRINT "K=";K;"RECIPROCAL ";A$
16Ø   END
RUN
ENTER HALF LIFE?162Ø
ENTER TIME UNITS - SECONDS, MINUTES, ETC.?YEARS
K= 4.27869E-Ø4   RECIPROCAL YEARS
DONE
```

This program is very straightforward. The LOG function in line 140 must be a
base e (natural) log; on some computers, both base e and base 10 (common) log functions are available.

It is often convenient to refer to a more directly measurable quantity than
the actual number of atoms. In radiochemistry, one frequently measures the radiation as the number of "counts per minute" with a Geiger counter, scintillation
counter, or other such device. Because the number of counts per minute (cpm) is
proportional to the number of radioactive atoms present, the decay equation can be
used with either units--atoms or cpm. Because the number of grams of a material is
proportional to the number of atoms present, grams can also be used.

Two very important applications of the radioactive decay equation are radiocarbon dating and neutron activation analysis.

Problem 11.1

The half-lives of a number of radioactive isotopes are given below. Use the
program in Example 11.1 to determine the decay constant for each material.

Isotope	Half-life
^{90}Sr	28 years
^{14}C	5600 years
^{19}O	29 seconds
^{235}U	7.1×10^{8} years
^{13}N	10 minutes
^{225}Ra	14.8 days

Problem 11.2

Assume that you start with a 10 g sample of each material listed in Problem 11.1. Write a program to prepare a neatly formatted table showing how much of each material would be left after 1 year, 5 years, and successive 10-year periods from 10 to 100 years.

11.2 Radiocarbon Dating

Radiocarbon dating is a method used in archaeology, paleontology, and other sciences in which one needs to find the age of objects containing organic matter. It has been found that naturally occurring radioactive carbon (^{14}C) is present in all plants due to the presence of $^{14}CO_2$ in the atmosphere. In living plants, the equilibrium ^{14}C level is such that these plants show a uniform radiation level of 15.3 counts per minute (cpm) per gram of carbon present. It is generally assumed that this equilibrium level has remained constant for at least the past 40,000 years. When the plant dies, this level of ^{14}C radioactivity begins to fall off according to the radioactive decay equation. Thus, the time elapsed since the death of a plant (and, consequently, the age of any object containing plant matter) can be found by measuring the current radiation level of the object and observing that

$$t \text{ (time elapsed)} = \frac{1}{k} \ln \frac{N_o}{N}$$

where the ratio N_o/N is equal to

$$\frac{15.3 \text{ cpm/g carbon}}{\text{current cpm/g carbon}}$$

The value of k is determined from the half-life of ^{14}C, which is 5,600 years.

Problem 11.3

Write a program to compute the radiocarbon age of any piece of wood given its current radioactivity level in cpm/g carbon. ^{14}C has a $t_{1/2}$ of 5,600 years. You can incorporate the program of Example 11.1 into your program.

Use your program to calculate the age of each of the following artifacts:

Artifact	Current count rate
piece of wooden sarcophagus from Egyptian tomb	7.6 cpm/g carbon
fragment of Dead Sea scroll	12.0 cpm/g carbon
wooden ax handle, American Indian	5.0 cpm/g carbon
remains of extinct animal	4.1 cpm/g carbon

Problem 11.4

Recent work based on counting growth rings in trees has shown that there are significant errors in radiocarbon dates for certain periods. What do you think is the weakest assumption in the radiocarbon method? (See "Carbon 14 and the Prehistory of Europe" by Colin Renfrew, Scientific American, 225:63 (October 1971).)

11.3 Neutron Activation Analysis

Imagine being given the world's only sample of a new inorganic compound, which took its discoverer six months to prepare. Imagine that there is only 0.001 g of the material in existence. Then, imagine being asked to determine the percentage of copper and bromine in the sample, without sacrificing any of the sample.

This seemingly impossible task can in fact be accomplished using neutron activation analysis (NAA). This method is selective, sensitive, and nondestructive. NAA can be used to analyze each of several elements present in a compound. The method can generally detect as little as 10^{-8} g of an element; and, with some elements, as little as 10^{-11} g can be detected. Finally, the sample is left intact by the analytical procedure.

NAA is based on the fact that most atoms become radioactive when exposed to a high-energy neutron flux in an atomic reactor. Once a particular element becomes radioactive through neutron absorption, its atoms decay with a characteristic emission. This emission will be of a type (γ-ray, neutron, α particle, etc.), energy, and half-life characteristic of that particular element. Thus, the elements present in an activated sample can be determined by measuring the types, energies, and half-lives of the radiations from the sample.

Because all radioactive emissions, regardless of type, follow the same decay equation, the actual amounts of the various elements present can be determined by measuring the emission rates of the various radiations. We recall that

$$N = N_o e^{-kt}$$

Suppose that the radiation from a sample is measured at time t_1 and is found to be N_1 cpm. From this equation, we can calculate N_o, the count rate immediately following irradiation. To convert N_o into the actual number of atoms of the element being analyzed, we would have to know:

1 Whether all the atoms of that element present in the sample had been rendered radioactive;

2 The frequency of emission of the atoms in question;

3 What fraction of the emissions actually reach the detector being used.

Because these data may not be available, one often uses a known reference sample. This would be a sample containing a known amount of the element in question. This reference sample would be irradiated simultaneously with the unknown sample. If the two samples were then counted simultaneously, we would simply have

$$\frac{\text{atoms of element X in unknown}}{\text{atoms of element X in reference}} = \frac{\text{cpm of unknown}}{\text{cpm of reference}}$$

The decay equation does not appear explicitly in this relation.

However, it is not always practical to count both samples simultaneously. In this case, we would measure the count rate of the unknown sample at time t_1 and the count rate of the reference sample at time t_2.
Then

$$N^{unk}(\text{cpm}) = N_o^{unk}(\text{cpm})\ e^{-kt_1}$$

$$N^{ref}(\text{cpm}) = N_o^{ref}(\text{cpm})\ e^{-kt_2}$$

Therefore,

$$\frac{\text{atoms of element X in unknown}}{\text{atoms of element X in reference}} = \frac{N_o^{unk}}{N_o^{ref}}$$

$$= \frac{N^{unk}\ e^{kt_1}}{N^{ref}\ e^{kt_2}} = \frac{N^{unk}}{N^{ref}}\ e^{k(t_1 - t_2)}$$

Thus, the amount of element X in the unknown sample can be calculated as

atoms of element X in unknown =

$$\left(\text{atoms of element X in reference}\right) \times \left(\frac{N^{unk}(\text{cpm})}{N^{ref}(\text{cpm})}\ e^{k(t_1 - t_2)}\right)$$

Because the number of grams of the element is proportional to the number of atoms, we could just as well use grams instead of atoms in this equation.

The formula just given assumes that k is known. Recall that k is related to the half-life of the emitting species as

$$k = \frac{\ln 2}{t_{1/2}}$$

Suppose, however, that the half-life of the emitting species is not known. This information can also be obtained from the neutron activation experiment. We observe that

$$N\,(cpm) = N_o\,(cpm)\,e^{-kt}$$

and

$$\ln N = \ln N_o - kt$$

We recognize this second expression as the equation of a straight line. Thus, if we measure N at various times t, and prepare a plot of ln N versus t, the points will lie on a straight line with slope equal to $(-k)$ and ordinate intercept equal to $\ln N_o$. Using the value of k so obtained, we can then determine $t_{1/2}$ and also proceed with the calculation of the amount of the element being analyzed.

Transforming an equation into linear form for the purpose of graphically determining some quantity is a fairly common trick in chemical circles. It is rather a clever gambit, don't you think? However, it leaves one with the problem of finding the best straight line through the experimental data points, because errors in measurement will cause the points to deviate from perfect linearity. The computer is very useful in dealing with this problem, which is treated in Chapter 14.

Problem 11.5

Suppose you are doing neutron activation activation analysis and have facilities for simultaneous counting of the unknown and reference samples. Write a general program that will determine the unknown amount of the element in question given the following INPUT data:

(a) count rate of the unknown sample (cpm)

(b) count rate of the reference sample (cpm)

(c) amount of the element present in the reference sample (g)

Use your program to help you in the following analysis:

It is necessary to identify the manufacturer of a fragment of a brass rifle cartridge casing found at the scene of a crime. Suppose the identification can be made by determining the percentage of copper in the brass fragment. The fragment weighs 0.01326 g. It is exposed in a nuclear reactor along with a reference sample, which is a 99.99 percent pure copper wire weighing 0.1506 g. After irradiation, the cartridge fragment and the reference sample are simultaneously counted. The fragment gives 88 cpm, while the reference sample gives 1,532 cpm. Determine the percentage copper in the cartridge case.*

Problem 11.6

Write a program for neutron activation analysis in the case where simultaneous counting of the unknown and reference samples is not possible. Your program will have to compute the unknown amount of the element in question given the amount of the element present in the reference sample, the count rate of the unknown sample and the time at which this count rate was measured, the count rate of the reference sample and the time at which its count rate was measured, and the half-life of the radioactive species.

Apply your program to the following problem:

A sample of "pure" iron is subjected to neutron activation. Detection of 2.8 MeV β^- particle emission from the irradiated sample indicates the presence of manganese as an impurity. The half-life of the radioactive species is known to be 2.55 hr. A reference sample is prepared consisting of 5.000 μg of Mn^{+2} in solution. This reference sample and a piece of the iron sample weighing 0.1759 g are simultaneously irradiated. These samples are then counted. The count rate of the iron sample is measured 10 min after removal from the reactor and is found to be 3,478 cpm. The manganese reference sample is counted 20 min after removal from the reactor and gives a count rate of 315 cpm. Find the percentage of manganese in the iron.

*NOTE: We are ignoring the fact that copper actually yields two different radio-active isotopes with different half-lives. For a discussion of this problem, see the article by Daly, Hofstetter, and Schmidt-Bleek in the Journal of Chemical Education, 44:412 (1967).

12

Problems in Molecular Structure Determination

Although no one has actually seen a molecule,* a chemist today can take a completely unknown compound and, within a month or two in favorable cases, determine the number and kind of the various atoms in its molecules, find the topological arrangement of these atoms (what is connected to what), establish the precise shape of the molecule, and even get a feel for the distribution of the electrons within the molecule.

This intimate structural information is gained by a variety of instrumental techniques. In some cases, the raw instrumental data must be processed by computer to yield physically interpretable information. The following pages will briefly describe some of these powerful instrumental techniques and propose computer experiments to illustrate their application. The reader will be referred to other sources for detailed discussions of the specific techniques.

*NOTE: This introductory clause is already false to some degree. It becomes increasingly false as research moves ahead. See, for example, the article by Ottensmeyer, Schmidt, and Olbrecht in Science, 179:175 (1973). It has been possible to obtain electron microscope images of certain special molecules. However, these images at the present time convey little in the way of detailed structural information.

12.1 Mass Spectrometry

A mass spectrometer breaks molecules into ionized fragments and measures the masses of the various fragments. The resulting data are presented as a bar graph of relative number of fragments versus mass, as shown in Figure 12.1.

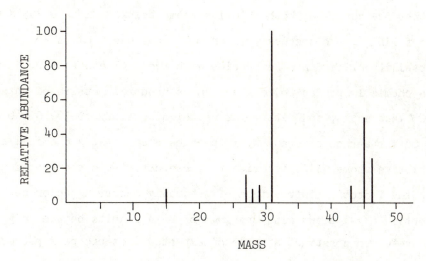

Figure 12.1   Schematic representation of the eight largest peaks in the mass spectrum of ethanol, CH_3CH_2OH.

This information has a number of uses. Many materials give rise to a mass spectral peak for the unfragmented or parent molecule. This peak, which would be the highest mass peak in the spectrum, directly gives the molecular weight of the compound. In working with mass spectra of organic compounds, the chemist also learns to recognize certain fragments such as $-CH_3$ (mass = 15 amu), $-CH_2CH_3$ (mass = 29 amu), etc. The occurrence of such species in the mass spectrum of an unknown material allows the chemist to reconstruct mentally the structure of the original molecule.

An interesting situation arises in the mass spectra of compounds containing halogens. Chlorine, for example, occurs in nature in two isotopic forms: ^{35}Cl and ^{37}Cl. Of all the chlorine atoms on earth, 75.5 percent are ^{35}Cl and 24.5 percent are ^{37}Cl; that is, there are about three times as many ^{35}Cl atoms as ^{37}Cl atoms. This means that if a sample of, say, CH_3Cl is analyzed in a mass spectrometer, there will be two parent peaks. There will be a peak at mass 50 corresponding to the CH_3Cl molecules that contain ^{35}Cl, and there will be a peak at mass 52 corresponding to the molecules containing ^{37}Cl. The mass 50 peak will be about three times as high as the mass 52 peak, because the two chlorine isotopes are present in nature in a 3:1 ratio. In fact, any molecular fragment containing one chlorine atom will show this characteristic pattern of two peaks in a 3:1 ratio, spaced by two mass units.

If we consider the case of a chemical species containing <u>two</u> chlorine atoms, we can see that the chance of both chlorine atoms being ^{35}Cl is 0.755 X 0.755; the probability of finding the species with one ^{35}Cl and one ^{37}Cl is 2 x 0.755 x 0.245; and the probability of finding the species with two ^{37}Cl atoms is 0.245 x 0.245. Thus, such a chemical species would give a mass spectral pattern of three peaks, spaced by two mass units, and with relative heights of 0.57:0.369:0.060 or 9.5:6.2:1. On further consideration, it becomes clear that a chemical species containing n chlorine atoms will give rise to a characteristic pattern of mass spectral peaks due to the isotopic distribution of naturally occurring chlorine. This pattern will contain n + 1 peaks with spacings of 2 mass units between adjacent peaks.

It is useful to construct a table of expected isotopic peak patterns for various numbers of chlorine atoms that might be present in a molecule or fragment. It will be seen after some reflection that a species containing n chlorine atoms will give n + 1 peaks whose relative heights (starting with the lowest mass peak in the pattern) will be:

$$\binom{n}{m} 0.755^{n-m} \times 0.245^{m}$$

for successive values of m from 0 to n.

$$\binom{n}{m}$$

is just the binomial coefficient and is equal to

$$\frac{n!}{(n-m)! \; m!}$$

Thus, for a three-chlorine species, there would be a pattern of four peaks with heights

$$\binom{3}{0} 0.755^{3} \times 0.245^{0}$$

$$\binom{3}{1} 0.755^{2} \times 0.245^{1}$$

$$\binom{3}{2} 0.755^{1} \times 0.245^{2}$$

$$\binom{3}{3} 0.755^{0} \times 0.245^{3}$$

or

1 x 0.430

3 x 0.140

3 x 0.0453

1 x 0.0147

It is conventional to list the largest peak in the group as having a height of 100 and to calculate the other peak heights relative to this one.   This group would then become

100

97.7

31.6

3.42

Problem 12.1

Write a program to prepare a table of expected isotope peak patterns for chemical species with 1 to 15 chlorine atoms.

Problem 12.2

Incorporate the following sequence into this program to produce actual simulated mass spectrometer patterns for such species.

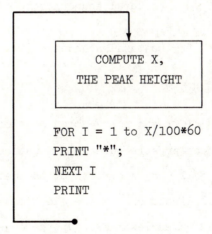

```
COMPUTE X,
THE PEAK HEIGHT

FOR I = 1 to X/100*60
PRINT "*";
NEXT I
PRINT
```

Figure 12.2   Program for mass spectrometer patterns.

Problem 12.3

A compound has the formula C_7Cl_8. In the mass spectrometer, this material gives the following sets of peaks:

Group 1.

Mass	117	119	121	123
Relative height	100	98	32	3

Group 2.

Mass	247	249	251	253	255	257
Relative height	62	100	65	21	3.4	0.22

Group 3.

Mass	364	366	368	370	372	374	376	378	380
Relative height	34	88	100	65	26	6.8	1.1	0.095	0.0041

Using your table from Problem 12.1 and the masses associated with the above peaks, determine the formula of each fragment and, if you can, the probable structure of the original compound.

Problem 12.4

Modify your program from Problem 12.1 to prepare a similar table of isotope peak patterns for bromine-containing compounds. Bromine has two naturally occurring isotopes

$$^{79}Br \quad \text{(natural abundance 50.6 percent)}$$

and

$$^{81}Br \quad \text{(natural abundance 49.4 percent)}$$

12.2 X-ray Diffraction

The methods of x-ray diffraction have been developed in recent years into a very powerful means of molecular structure determination. In this technique, a beam of x-rays is made to fall upon a crystal of the material under study. The x-rays are scattered in all directions by the electrons in the molecules. Because of the orderly array of the molecules making up the crystal, these scattered x-rays interfere, and detectable x-rays leave the crystal only in certain directions. The intensities and directions of these scattered beams depend on the arrangement of the atoms in the molecules (molecular structure) and on the arrangement of the molecules in the crystal (crystal structure). Several steps of computer calculation are necessary to convert the measurements of the scattered x-rays into physically meaningful data. The final result is a list of the identities and the precise positions of the atoms in the "unit cell" of the crystal. These positions are given as a set of three-dimensional coordinates of each atom in space.

This list of atomic positions, however, does not itself give the molecular structure. One needs to know how the atoms are connected together. This is discovered by computing the distances between the various pairs of atoms, using the list of atomic coordinates. These interatomic distances are then compared with the known lengths of the various types of chemical bonds. For example, if two carbon atoms in the crystal are found to be 1.54Å apart, then we would infer that there must be a single bond between them, because C-C single bonds are always found to have lengths close to this value. In this way, it is possible to determine which pairs of atoms are connected together, and ultimately the structure of the whole molecule can be drawn.

Suppose two atoms have coordinates (x_1, y_1, z_1) and (x_2, y_2, z_2). The distance between the atoms is given by

$$d = ((x_1 - x_2)^2 + (y_1 - y_2)^2 + (z_1 - z_2)^2)^{1/2}$$

In addition, if the coordinates of three atoms are known, one can calculate the angles among them from the three interatomic distances. In Figure 12.3, three atoms are depicted, with interatomic distances d_{12}, d_{13}, and d_{23}.

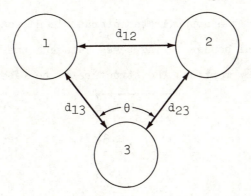

Figure 12.3 Relation between inter-bond angle and interatomic distances.

It can be shown that

$$\cos \theta = \cos \angle 132 = \frac{d_{13}^2 + d_{23}^2 - d_{12}^2}{2\, d_{13}\, d_{23}}$$

The angle itself cannot be computed directly from this relation unless your computer has an inverse cosine predefined function. In the case that only the inverse tangent function (ATN) is available, one can still extract the angle θ by observing that

$$\tan^2\theta = \frac{\sin^2\theta}{\cos^2\theta} = \frac{1 - \cos^2\theta}{\cos^2\theta}$$

so that

$$\theta = \arctan \sqrt{(1 - \cos^2\theta)/\cos^2\theta}$$

Problem 12.5

Write a program to accept the coordinates of two atoms and to compute the distance between them.

Problem 12.6

Use the program from Problem 12.5 to compute the distances between the following atoms in a compound with the formula CH_4ON_2.

Atom	x, Å	y, Å	z, Å
C	0.000	2.831	1.560
N1	0.8034	2.028	0.8758
N2	-0.8034	3.634	0.8758
O	0.0000	2.831	2.829

Problem 12.7

Here is a table of the approximate interatomic distances associated with various kinds of chemical bonds. Using these data and the interatomic distances computed in Problem 12.6, work out the structure of the above compound.[*]

	d, Å
C-C	1.54
C=C	1.34
C≡C	1.20
C-O	1.43
C=O	1.23
C-N	1.47
C=N	1.3
C-H	1.1
O-H	1.0

[*]For a discussion of the structure of this molecule, CH_4ON_2, see A. F. Wells, "Structural Inorganic Chemistry," 3rd edition, Oxford University Press, 1962, p. 708.

Problem 12.8

Write a program to accept the coordinates of three atoms and to compute the bond angle. Take the middle atom as the vertex of the angle. Remember that the computer deals with angles in radians (see Chapter 6). Also, beware of the fact that the function given here for evaluating θ using the ATN function will give a DIVIDE-BY-ZERO error message if $\cos \theta = 0$; this will happen if $\theta = 90^\circ$. To avoid this pitfall, add a very small number (say, $1.E-10$) to the denominator so that the denominator is never actually zero. This ploy, of course, introduces a slight error into the calculation, and one must choose the number added to the denominator to be small enough that the error is not significant.

Problem 12.9

Modify the program from Problem 12.5 to accept the coordinates of a _list_ of atoms, to calculate the interatomic distance between each pair of atoms, and to print out those pairs of atoms that are separated by distances less than, say, $2\overset{\circ}{A}$. The output should also include the actual distance between those pairs that are printed out. A list of this type is very useful in determining which atoms in a structure are actually bonded together. (Hint: read the atomic coordinates into three arrays: $X(I)$, $Y(I)$, $Z(I)$.) Test your program with the data from Problem 12.6.

12.3   Electron Diffraction

X-ray diffraction analysis, as described in the preceding section, generally requires that the material being studied be crystalline. Electron diffraction, in contrast, is a technique used to study the molecular structure of materials that are gases. A beam of electrons is directed through the gas sample. Diffraction causes the electrons to be scattered in a concentric circular pattern of graduated intensity which is recorded on film. After correction for background darkening of the film, this pattern is usually represented as a graph of intensity versus scattering angle θ, as shown in Figure 12.4.

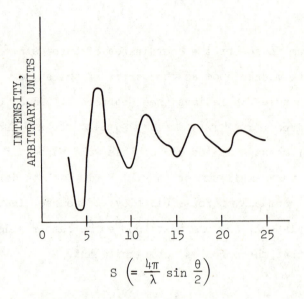

$$S \left(= \frac{4\pi}{\lambda} \sin \frac{\theta}{2}\right)$$

Figure 12.4 Graph of electron diffraction pattern of CO_2. Data from I. L. Karle
and J. Karle, Journal of Chemical Physics, 17:1052 (1949).

For a gaseous sample, we cannot discover the absolute positions of atoms,
because the molecules are in constant random motion. We can, nevertheless, find
the interatomic distances. This can be done either by a technique known as Fourier
inversion (one of the computational steps in x-ray diffraction analysis) or by calculating a theoretical electron scattering pattern based on a model of the molecule
and then varying the proposed shape of the model to give the best agreement between
the theoretical and actual scattering patterns. In this latter method, the basis
for calculating the theoretical scattering patterns is the Wierl equation. In its
most general form, the Wierl equation gives the diffracted intensity as a function
of the atomic number of the atoms in the molecule, the distances between each pair
of atoms in the molecule, and the diffraction angle.

The Wierl equation is

$$I(\theta) \propto \sum_{\substack{j \\ k \\ j \le k}} Z_j Z_k \frac{\sin\left[\frac{4\pi}{\lambda} \sin\left(\frac{\theta}{2}\right) r_{jk}\right]}{\frac{4\pi}{\lambda} \sin\left(\frac{\theta}{2}\right) r_{jk}}$$

where the summation runs over all atoms j and k, Z_j is the atomic number of atom j,
r_{jk} is the distance between atoms j and k, and λ is the wavelength of the electron
beam. If we define a new quantity,

$$s = \frac{4\pi}{\lambda} \sin \frac{\theta}{2}$$

then the equation takes on a much simpler appearance:

$$I(s) \propto \sum_{\substack{j \\ j \leq k}} \sum_{k} Z_j Z_k \frac{\sin(sr_{jk})}{sr_{jk}}$$

For example, for the hypothetical molecule

the Wierl equation would be

$$I(s) \propto Z_A^2 + Z_B^2 + Z_C^2 + Z_A Z_B \frac{\sin(sr_{AB})}{sr_{AB}} + Z_B Z_C \frac{\sin(sr_{BC})}{sr_{BC}} + Z_A Z_C \frac{\sin(sr_{AC})}{sr_{AC}}$$

Problem 12.10

One way of utilizing electron diffraction data is by matching the observed scattering pattern against theoretical scattering patterns calculated from the Wierl equation using various trial values for the molecular dimensions.

Consider the CO_2 molecule. Assuming the molecule to be linear and symmetrical, write out the Wierl equation (as a function of s, not θ) for CO_2. Then write a program to produce a simulated electron diffraction curve for CO_2 using an INPUT value for the C-O bond distance. Utilize the plotting routine given in Example 8.2 (Chapter 8). Run your program for values of the bond distance between 1.0 and 1.5 Å in intervals of 0.1 Å. Which of these theoretical curves most closely approximates the actual curve shown in Figure 12.4? Rerun your program for four or five values of the bond length close to the value just found. Try to determine the bond length to a precision of 0.02Å. Compare your value with values found in the chemical literature. (Consult, for example, I. L. Karle and J. Karle, _Journal of Chemical Physics_, _17_:1052 (1949).)

Problem 12.11

The above approach to electron diffraction analysis assumes a prior knowledge of the general architecture of the molecule being studied. If the basic structure is unknown, information can still be extracted from the diffraction pattern. The Fourier method allows us to compute a new function from the experimental scattering curve (Figure 12.4). The new function is

$$D(r) = \sum_i I_i \frac{\sin(s_i r)}{s_i r}$$

where I_i is the intensity of the scattering curve (Figure 12.4) at point i and s_i is the s value at this point. The function $D(r)$ is called the radial distribution function; this function has peaks at values of r which correspond to interatomic distances in the sample being studied. Thus, the values of the interatomic distances can simply be read off the $D(r)$ plot.

Write a program to evaluate and plot $D(r)$ given a set of I_i's and the corresponding s_i's. Utilize the plotting routine in Example 8.3 (Chapter 8). With your program, plot $D(r)$ for CO_2 from r = 0 to $3\overset{o}{A}$ in increments of $0.1\overset{o}{A}$. Obtain the necessary values of I_i and s_i from the CO_2 scattering curve in Figure 12.4. Use the I and s values corresponding to the four peaks on that curve. (The intensity scale in Figure 12.4 is arbitrary; the Y-axis scaling limits for your $D(r)$ plot will depend on how you define the intensity scale for Figure 12.4.)

Examine the resulting radial distribution curve. What value do you obtain for the C-O bond length? What is the significance of the second peak in the radial distribution curve?

Problem 12.12

Using the program created in Problem 12.11, run $D(r)$ plots for CO_2 utilizing various amounts of experimental data. That is, run $D(r)$ plots using only one, then two, and then four of the peaks in Figure 12.4 as the I_i values in your program. Then run two more $D(r)$ plots making use of 20 data points and then 40 data points taken from Figure 12.4. (Do not select values of s less than 3, however.) Compare these radial distribution curves made with various amounts of experimental data. Pay particular attention to the resulting values for the C-O and O-O bond lengths, as well as to the presence of small, spurious peaks.

What do you conclude about the need for experimental data in the Fourier method? For a more complex molecule, what do you think would happen to the need for data?

12.4 Nuclear Magnetic Resonance Spectrometry

Percy Bridgman once defined science as "doing your damndest with your mind, and no holds barred." When it comes to ferreting out information about the structures of molecules, the chemist indeed bars no holds. Any molecular property (however obscure) that is sensitive to variations in structure gives a potential "handle" on the structure itself and is fair game for structural studies. The phenomenon of nuclear magnetic resonance (NMR) is a rather subtle effect, which requires sophisticated electronic equipment merely to detect it. Nonetheless, NMR has evolved into a powerful and widely used tool for structure elucidation.

Atomic nuclei in which the number of protons and the number of neutrons are not both even possess a nonzero magnetic moment. This means that the nuclei behave like magnets. If a sample containing such nuclei is placed in a magnetic field, two possible energy states are established for the nuclei:

1 The nuclear "magnets" can be lined up parallel with the applied field;

2 The nuclear "magnets" can be lined up antiparallel with the applied field. Now, ordinarily, the majority of these tiny nuclear "magnets" will line up parallel with the imposed field, because this represents the lowest energy orientation. However, if a radio frequency (r.f.) field of exactly the right frequency (energy) is introduced, the nuclei will absorb energy from the r.f. field and will flip to the higher energy antiparallel orientation with respect to the applied field. The difference in energy between the two orientations and, thus, the exact frequency at which this absorption or resonance occurs is proportional to the strength of the magnetic field experienced by the nuclei.

What has this to do with molecular structure? The magnetic field experienced by the nuclei determines the absorption frequency. This magnetic field is established, of course, by an external magnet, but the field actually experienced by any nucleus is modified by the "chemical environment" of that nucleus. Specifically, the electrons surrounding any nucleus shield it from the external field. A hydrogen nucleus in an H_2 molecule is shielded by the electrons of the H-H bond. If one of those H atoms were now attached instead to a more electronegative atom, say, carbon, the H nucleus would have a smaller share of the electron pair and thus would be more exposed to the external field. As a result, the NMR absorption frequency would be shifted. If the carbon atom were then replaced by a more electronegative oxygen atom, the H nucleus would become even more deshielded, and the absorption frequency would be shifted yet further. Thus, the NMR absorption frequency of a nucleus depends on its chemical surroundings, that is, on the structure of the molecule in which the nucleus resides.

The displacement of the absorption frequency is called the "chemical shift". The chemical shift is usually measured relative to the magnetic resonance absorption frequency for a standard substance, often tetramethylsilane, $(CH_3)_4 Si$. Note that all 12 hydrogen atoms in tetramethylsilane (TMS) are in identical chemical environments. If the absorption frequency of the hydrogen nuclei in TMS is arbitrarily assigned to be 0.0 Hz ("Hertz" or cycles per second), then we would find, for example, that the absorption frequency of the hydrogen nuclei in $CH_3 Cl$, which are less shielded than those in TMS, would be 180 Hz higher than that of the TMS protons. (The size of the chemical shift also depends on the field strength of the external magnet. The standard field strength is 14,092 gauss, and the shifts quoted here are based on that field strength.)

Although many different nuclei will give NMR absorptions, it is the hydrogen nuclei that are most often studied. The most common NMR instruments are designed to produce hydrogen (proton magnetic resonance) spectra.

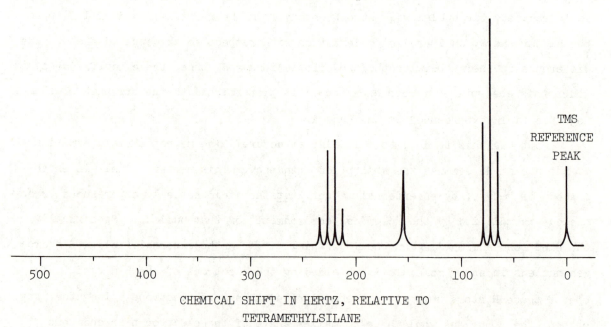

CHEMICAL SHIFT IN HERTZ, RELATIVE TO
TETRAMETHYLSILANE

Figure 12.5 NMR spectrum of ethanol, $CH_3 CH_2 OH$, dissolved in carbon tetrachloride.

If the chemical shift were the only effect operating, a PMR (proton magnetic resonance) spectrum of a molecule would consist of one line for each kind of chemical environment in which H atoms occur in the molecule. As you can see from Figure 12.5, the real situation is more complex. Although hydrogen atoms reside in only 3 different chemical environments in ethanol, CH_3CH_2OH, the PMR spectrum has 8 peaks. This is due to a phenomenon called spin-spin splitting. The explanation of spin-spin splitting is not difficult, but it will not be given here.* Suffice it to say that usually the absorption line for a group of chemically identical protons attached to a given atom will be split into $n+1$ sublines, where n is the number of protons on the immediately adjacent atoms. The spacing between the split sublines is given by the "coupling constant", which is measured in Hz and is characteristic for the particular structural situation.

To get a feel for the relation between molecular structure and PMR spectrum appearance, one should ideally sit down at the NMR spectrometer with a group of wisely chosen samples and study the spectra to see the influence of differences in molecular structure. As an alternative, one can study the excellent spectra collected in a book such as the "Varian NMR Spectra Catalog," Volumes 1 and 2, Varian Associates, 1962 and 1963.

However, the available samples or the available spectra are not always appropriate to the question in one's mind. In such an instance, it is useful to be able to simulate the NMR spectrum that would arise from molecule X if it were available. The calculation of an NMR spectrum, given the structure of a molecule, is a very big computational job in all but fairly simple cases. Computers, of course, are good at big computational jobs. Using the computer (and a sophisticated program), it is possible to simulate very complex NMR spectra.** In fact, such simulations are sometimes used in reverse, to help figure out the structure of a new molecule by trying to think up a structure whose computer-generated spectrum matches the actual NMR spectrum of the molecule.

---

*For a more detailed discussion of NMR spectra, see G. Barrow, "Physical Chemistry," McGraw-Hill Book Company, 1973; and F. A. Bovey, "Nuclear Magnetic Resonance," Academic Press, 1969.

**See, for example, Kenneth B. Wiberg and Bernard J. Nist, "The Interpretation of NMR Spectra," W. A. Benjamin, Inc., New York, 1962.

Problem 12.13

The following program computes PMR spectra in a simple situation,

$$R-\overset{|}{\underset{A}{C}}-\overset{|}{\underset{B}{C}}-R'$$

Here, we have two hydrogen atoms (A and B) on adjacent carbon atoms which do not
constitute identical chemical environments. The <u>difference</u> in the chemical shifts
of the two environments is INPUT to the program, as is the coupling constant for
the spin-spin coupling between the two hydrogens.

Using this program, investigate how the spectrum varies with changing conditions. Observe the relative peak heights as well as the peak positions. Typical
coupling constants for hydrogen atoms on adjacent carbons are in the range of 2 to
9 Hz. Here are some suggestions:

(a) What does the spectrum look like when both protons are in identical
     environments (have the same chemical shift)? The coupling constant (J)
     will also have to be zero in this case.

(b) For J = 0, what happens as the difference in chemical shift becomes
     greater?

(c) For a given difference in chemical shift, what happens as the coupling
     constant is increased?

(d) For a given nonzero coupling constant, what happens as the chemical
     shift difference increases?

(e) What happens when the chemical shift difference is quite large compared
     with the coupling constant?

Program to generate PMR spectra for the AB case. (This program is reproduced with
the kind permission of its creator, Dr. James U. Piper, Simmons College, Boston,
Mass.)

```
10  PRINT "THIS PROGRAM COMPUTES THE POSITIONS OF THE FOUR LINES"
15  PRINT "OF AN AB SPECTRUM CENTERED AT 50 HZ.  IT ALSO COMPUTES"
20  PRINT "THE RATIO OF THE INTENSITIES OF LINES 1 AND 2 (OR 4 "
25  PRINT "AND 3).  IT THEN PRINTS THE APPROXIMATE SPECTRUM."
30  PRINT "WHAT ARE THE VALUES FOR THE CHEMICAL SHIFT DIFFERENCE"
31  PRINT "(F) AND THE COUPLING CONSTANT (J) IN HZ?"
```

```
35    PRINT

4Ø    INPUT F,J

5Ø    PRINT "F=";F;"HZ","J=";J;"HZ"

51    PRINT

52    PRINT

55    LET F=ABS(F)

56    LET J=ABS(J)

6Ø    LET X=SQR(F↑2 + J↑2)

65    REM COMPUTE RELATIVE INTENSITIES

7Ø    LET I4=1-J/X

75    LET I3=1+J/X

77    REM COMPUTE LINE POSITIONS

8Ø    LET M=5Ø-(X+J)/2

85    LET N=5Ø-(X-J)/2

9Ø    LET P=5Ø+(X-J)/2

95    LET Q=5Ø+(X+J)/2

1ØØ   PRINT "LINE 1 OCCURS AT";M;"HZ"

1Ø5   PRINT "LINE 2 OCCURS AT";N;"HZ"

11Ø   PRINT "LINE 3 OCCURS AT";P;"HZ"

115   PRINT "LINE 4 OCCURS AT";Q;"HZ"

124   PRINT "THE RATIO OF INTENSITIES 1/2 OR 4/3 =";I4/I3

125   PRINT

126   PRINT

13Ø   REM PLOTTING SECTION

135   FOR B=INT(M+.5) TO INT(Q+.5)

14Ø   IF INT(B/1Ø)=B/1Ø THEN 17Ø

15Ø   LET A$="."

16Ø   GOTO 175

17Ø   LET A$="-"

175   PRINT A$;

18Ø   IF B#INT(M+.5) THEN 23Ø

19Ø   LET I1=INT(3Ø*(1-J/X)+.5)

2ØØ   FOR I=1 TO I1
```

```
21Ø    PRINT "*";
22Ø    NEXT I
23Ø    IF B#INT(N+.5) THEN 28Ø
24Ø    LET I2=INT(3Ø*(1+J/X)+.5)
25Ø    FOR I=1 TO I2
26Ø    PRINT "*";
27Ø    NEXT I
28Ø    IF B#INT(P+.5) THEN 33Ø
29Ø    LET I2=INT(3Ø*(1+J/X)+.5)
3ØØ    FOR I=1 TO I2
31Ø    PRINT "*";
32Ø    NEXT I
33Ø    IF B#INT(Q+.5) THEN 38Ø
34Ø    LET I1=INT(3Ø*(1-J/X)+.5)
35Ø    FOR I=1 TO I1
36Ø    PRINT "*";
37Ø    NEXT I
38Ø    PRINT
39Ø    NEXT B
391    PRINT
392    PRINT
393    PRINT
4ØØ    END
RUN
```

THIS PROGRAM COMPUTES THE POSITIONS OF THE FOUR LINES
OF AN AB SPECTRUM CENTERED AT 5Ø HZ.   IT ALSO COMPUTES
THE RATIO OF THE INTENSITIES OF LINES 1 AND 2 (OR 4
AND 3).   IT THEN PRINTS THE APPROXIMATE SPECTRUM.
WHAT ARE THE VALUES FOR THE CHEMICAL SHIFT DIFFERENCE
(F) AND THE COUPLING CONSTANT (J) IN HZ?
?1Ø,2
F= 1Ø   HZ   J= 2   HZ

```
LINE 1 OCCURS AT 43.901      HZ
LINE 2 OCCURS AT 45.901      HZ
LINE 3 OCCURS AT 54.099      HZ
LINE 4 OCCURS AT 56.099      HZ
THE RATIO OF INTENSITIES 1/2 OR 4/3 = .672078


.*************************
.
.************************************
.
.
.
-
.
.
.
.*******************************************
.
.***************************
DONE
```

For further reading in molecular structure determination:

1 General: "Physical Chemistry," 3rd edition, Gordon Barrow, McGraw-Hill Co.,
                 New York, 1973.

2 Mass spectrometry: "Introductory Mass Spectrometry," Stephen R. Shrader,
                           Allyn and Bacon, Inc., Boston, 1971.

3 X-ray diffraction: "X-ray Structure Determination," George H. Stout and
                           Lyle H. Jensen, Macmillan Co., New York, 1968.

4 Electron diffraction: "The Determination of Molecular Structure," 2nd
                              edition, P. J. Wheatley, Oxford University
                              Press, Oxford, 1968.

5 Nuclear magnetic resonance: "Nuclear Magnetic Resonance," Frank A. Bovey,
                                    Academic Press, New York, 1969.

13

Atomic Theory and Chemical Bonding

The wave mechanical approach has become an increasingly important route to the understanding of atomic and molecular structure and behavior. It is fair to say that, without computers, this road to understanding and insight would be at best a narrow, unpaved path. The basic formulations of wave mechanics when applied to systems of reasonable complexity--polyelectronic atoms and molecules--lead to equations that defy solution. To deal with these equations even approximately, using analytic methods, has demanded the innovative efforts of the best mathematicians and physicists.

With the advent of the digital computer, the treatment of difficult equations by "numerical" rather than by analytic methods became a realistic prospect. Numerical methods involve a systematic search for actual, numerical values that satisfy a specific equation, while the analytic approach seeks a formula that solves a general equation. Typically, a numerical attack on a problem involves testing thousands or millions of trial solutions, gradually converging on the best solution at a specified level of precision. Clearly, only the computer can handle the computational demands of such a method. This strategy may offend one's sense of mathematical elegance, but it is proving increasingly successful in yielding hard, a priori atomic and molecular data such as structures and dimensions of molecules, electron distributions and, by implication, chemical reactivities.

214

The computer programs used in wave mechanical research tend to be large, sophisticated, and long-running.[*] It is not unusual for such work to consume an hour at a time on the largest, fastest, computers available (at, say $500/hr). The examples that follow do not approach this level but will give you an idea of some of the types of things the computer can do in the area of theoretical atomic and molecular investigation.

Problem 13.1

Certain properties of chemical bonds can be understood in terms of the "percent ionic character" of the bonds. This concept recognizes the fact that a bond is rarely totally ionic or totally covalent.

Linus Pauling suggested that the percent ionic character of a bond between two atoms could be estimated by comparing the electronegativities of the atoms participating in the bond. Pauling assigned each element an electronegativity on an arbitrary scale. The difference in the electronegativities of two atoms correlates with the ionic character of a bond between them, as shown in Table 13.1.

Table 13.1 Percent ionic character versus difference in electronegativity.

Difference in electronegativity of the atoms in the bond	Percent Ionic Character of the bond
0.0	0
0.1	0.5
0.2	1
0.3	2
0.4	4
0.5	6
0.6	9
0.7	12
0.8	15
0.9	19
1.0	22

[*]See, for example, Arnold C. Wahl, "Chemistry by Computer," _Scientific American_, _222_:54 (April 1970).

1.1	26
1.2	30
1.3	34
1.4	39
1.5	43
1.6	47
1.7	51
1.8	55
1.9	59
2.0	63
2.1	67
2.2	70
2.3	74
2.4	76
2.5	79
2.6	82
2.7	84
2.8	86
2.9	88
3.0	89
3.1	91
3.2	92

Table 13.2 gives the Pauling electronegativities of the first twenty elements.

Table 13.2 Pauling electronegativities.

Element	Electronegativity
H	2.1
He	-
Li	1.0
Be	1.5
B	2.0
C	2.5
N	3.0
O	3.5
F	4.0
Ne	-
Na	0.9
Mg	1.2
Al	1.5
Si	1.8
P	2.1
S	2.5
Cl	3.0
Ar	-
K	0.8
Ca	1.0

Write a program incorporating information retrieval features (see Chapter 7) to accept as input any pair of elements from the first 20 elements and to give in return the percent ionic character of a bond between atoms of those elements. [Hint: store the listed electronegativity differences in one array, $A(I)$, and the corresponding percent ionic characters in a parallel array, $B(I)$.]

Use your program to find the approximate ionic character of the bond in each of the following compounds:

$$KF \quad NaF \quad LiF \quad H_2O \quad H_2S \quad H_2Se \quad CO_2 \quad SO_2 \quad Cl_2$$

Problem 13.2

The previous project uses table-look-up to find the percent ionic character of a chemical bond, given the electronegativities of the atoms participating in the bond. An analytic expression has been proposed to relate electronegativity differences to percent ionic character. The formula is

$$\text{percent ionic character} = 100 \, (0.16D + 0.035D^2)$$

where D is the electronegativity difference.

Write a program that accepts as input any pair of elements from the first 20 elements and uses the above formula to give in return the percent ionic character of a bond between atoms of the two elements. Use your program to calculate the ionic character of the molecules listed in Problem 13.1. Compare the results of the two programs. Is there a noticeable difference in the execution time of the two programs on your computer?

Problem 13.3

The following program calculates and plots either the wave function or the radial electron density function for the electron in a hydrogenlike (one-electron) atom. Place the program on your computer and use it to study these items:

(a) Compare the wave function and the radial electron density function for a 1s electron ($n = 1$, $\ell = 0$) and a nuclear charge of $+1$ ($Z = 1$).

(b) Do the same for a 2s electron ($n = 2$, $\ell = 0$) and a nuclear charge of $+1$.

(c) Compare the radial electron density function of a 2s electron with that of a 2p electron ($n = 2$, $\ell = 1$), both with a nuclear charge of $+1$.

(d) If the radius of an atom is taken to be approximately equal to the distance from the nucleus to the largest peak in the radial electron density function, use the program to find the approximate radius of a hydrogen atom.

(e) Compare the approximate radius of a 1s electron shell with the radii of a 2s electron shell and a 3s electron shell, all for atoms with the same nuclear charge.

(f) Compare the approximate radius of a 2s electron shell in a hydrogenlike atom with nuclear charge of a +1 with the radius of a 2s electron shell in an atom with nuclear charge of +2.

(Note: A good range for these plots is from 0 to 10Å in increments of 0.25Å. Test how changing these values affects the appearance of the plots.)

```
5Ø    REM PROGRAM FOR PLOTTING HYDROGENLIKE RADIAL WAVEFUNCTIONS
52    REM      OR ELECTRON DENSITY FUNCTIONS
6Ø    LET W=1
65    DIM A$[1Ø]
7Ø    PRINT "ENTER 'WAVE' FOR WAVEFUNCTION"
72    PRINT "  OR 'DENS' FOR ELECTRON DENSITY FUNCTION"
74    INPUT A$
76    IF A$="WAVE" THEN 9Ø
78    LET W=2
9Ø    PRINT
1ØØ   REM LIMITS ON INDEP. VARIABLE
1Ø5   REM X REPRESENTS DISTANCE FROM NUCLEUS IN ANGSTROMS
11Ø   PRINT "ENTER XMIN";
12Ø   INPUT X2
13Ø   PRINT "ENTER XMAX";
14Ø   INPUT X3
15Ø   PRINT "ENTER XINCR";
16Ø   INPUT X1
17Ø   PRINT
18Ø   REM INTEGERIZING FUNCTIONS
19Ø   DEF FNI(X)=INT(X/X1+.5)
2ØØ   DEF FNJ(Y)=INT((Y-Y2)/D+.5)+1
299   REM  Z=NUCLEAR CHARGE   N,L=PRINCIPAL,AZIMUTHAL QUANTUM #'S
```

```
300    PRINT "ENTER Z,N,L"

310    PRINT "Z=";

320    INPUT Z

330    PRINT "N=";

340    INPUT N

350    PRINT "L=";

360    INPUT L1

365    PRINT

370    LET A1=N+L1

380    LET B1=2*L1+1

390    DEF FNS(X)=(Z*X)/.529

405    REM RADIAL WAVEFUNCTION DEFINITION

410    DEF FNA(X)=D9*((X9+.000001)↑L1)*(EXP(-FNS(X)/N)*T)

415    REM ELECTRON DENSITY FUNCTION DEFINITION

420    DEF FNB(X)=((FNA(X))↑2)*(X↑2)

425    REM FACTORIAL ROUTINES FOR (N-L-1)! AND (N+L)!

430    LET N2=(N-L1)-1

440    LET K6=1

450    IF N2=0 THEN 490

460    FOR I=1 TO N2

470    LET K6=K6*I

480    NEXT I

490    LET N3=N+L1

500    LET K8=1

510    IF N3=0 THEN 550

520    FOR I=1 TO N3

530    LET K8=K8*I

540    NEXT I

550    LET C9=-SQR((4*K6)/((N↑4)*(K8↑3)))

560    LET C9=C9*(-1)↑(N+L1)*K8/K6

570    LET D9=C9*((Z/.529)↑1.5)

580    PRINT "AUTOMATIC SCALING";
```

```
582   INPUT A$
584   PRINT
586   IF A$="NO" THEN 85Ø
6ØØ   REM AUTO SCALING SECTION
61Ø   X=X2
62Ø   GOSUB 8ØØØ
63Ø   GOSUB W OF 81ØØ,82ØØ
7ØØ   LET C1=C2=Y
71Ø   FOR X=X2 TO X3 STEP X1
72Ø   GOSUB 8ØØØ
73Ø   GOSUB W OF 81ØØ,82ØØ
75Ø   IF C1 <= Y THEN 77Ø
76Ø   LET C1=Y
77Ø   IF C2 >= Y THEN 79Ø
78Ø   LET C2=Y
79Ø   NEXT X
8ØØ   PRINT "YMIN= ";C1
81Ø   PRINT "YMAX= ";C2
82Ø   LET Y2=C1
83Ø   LET Y3=C2
84Ø   GOTO 9ØØ
85Ø   REM LIMITS ON DEP. VARIABLE
855   PRINT
86Ø   PRINT "ENTER YMIN";
87Ø   INPUT Y2
88Ø   PRINT "ENTER YMAX";
89Ø   INPUT Y3
9ØØ   LET D=(Y3-Y2)/7Ø
91Ø   PRINT "YINCR= ";D
99Ø   GOSUB 9ØØØ
1ØØØ  REM PLOT GENERATOR
1ØØ5  FOR X=X2 TO X3 STEP X1
1Ø1Ø  GOSUB 8ØØØ
```

```
1Ø15    GOSUB W OF 81ØØ,82ØØ

1Ø2Ø    LET Y=FNJ(Y)

1Ø3Ø    PRINT ".";

1Ø4Ø    IF Y>71 THEN 1Ø7Ø

1Ø5Ø    IF Y<1 THEN 1Ø7Ø

1Ø58    REM PRINTS POINT

1Ø6Ø    PRINT TAB(Y);"*";

1Ø7Ø    PRINT

1Ø8Ø    NEXT X

2ØØØ    GOSUB 9ØØØ

7998    REM SUBROUTINE FOR ASSOCIATED LAGUERRE POLYNOMIAL

7999    STOP

8ØØØ    LET X9=FNS(X)*2/N

8Ø1Ø    LET T=Ø

8Ø2Ø    LET C3=1

8Ø3Ø    FOR N1=(A1-B1) TO Ø STEP -1

8Ø4Ø    LET T=T+C3*(X9+.ØØØØØ1)↑N1

8Ø5Ø    LET C3=C3*N1*(N1+B1)/(N1+B1-A1-1+.ØØØØØ1)

8Ø6Ø    NEXT N1

8Ø8Ø    RETURN

8Ø98    REM EVALUATE WAVEFUNCTION

81ØØ    LET Y=FNA(X)

811Ø    RETURN

8198    REM EVALUATE ELECTRON DENSITY FUNCTION

82ØØ    LET Y=FNB(X)

821Ø    RETURN

8997    REM AXES SUBROUTINE

8999    STOP

9ØØØ    PRINT " ";

9Ø1Ø    FOR I=1 TO 71

9Ø2Ø    PRINT ".";

9Ø3Ø    NEXT I

9Ø4Ø    PRINT
```

```
9Ø5Ø  RETURN
9999  END
RUN
ENTER 'WAVE' FOR WAVEFUNCTION
  OR 'DENS' FOR ELECTRON DENSITY FUNCTION
?WAVE
ENTER XMIN?Ø
ENTER XMAX?1Ø
ENTER XINCR?.25
ENTER Z,N,L
Z=?1
N=?2
L=?Ø
AUTOMATIC SCALING?YES
YMIN= -.247112
YMAX=  1.83781
YINCR=  2.97847E-Ø2
```

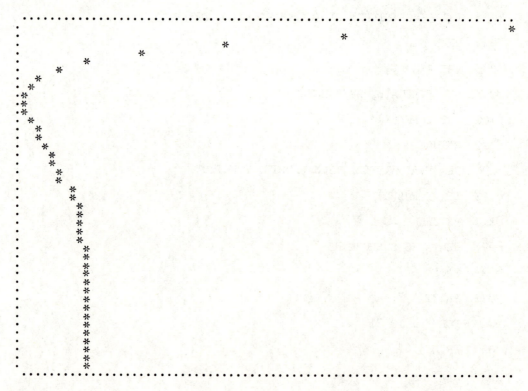

```
DONE
```

Problem 13.4

The interpretation of the emission spectrum of atomic hydrogen was a turning point in the development of atomic theory. Rydberg gave a formula that compactly summarized the wavelengths, λ, of these spectral lines:

$$\frac{1}{\lambda} = R\left(\frac{1}{n_1^2} - \frac{1}{n_2^2}\right)$$

where n_1 and n_2 are integers with $n_2 > n_1$ and $R = 1.096776 \times 10^{-3} \overset{\circ}{A}{}^{-1}$. Bohr later interpreted this formula on the basis of an atomic model in which n_2 is the quantum orbit number of the orbit from which an electron falls and n_1 is the quantum orbit number of the orbit to which the electron falls.

Write a program to evaluate λ, given n_1 and n_2 as input values. Use your program to help you determine which electron shells are involved in the transition giving rise to each of the following hydrogen spectral lines (in Ångstroms):

1215.68
1025.73
 972.54

6562.79
4861.33
4340.47
4101.74
3970.07

40500
26300
18751
12817

Problem 13.5

One must bear in mind that the computer, being a digital device, works with a limited number of significant figures. As the computer goes through a series of calculations, various truncations and roundings off occur, which, in a lengthy computation, may seriously affect the accuracy of the final result.

The following formula is mathematically equivalent with the one given in Problem 13.4:

$$\lambda = \frac{1}{R}\left(\frac{n_1^2 \, n_2^2}{n_2^2 - n_1^2}\right)$$

Modify your program from Problem 13.4 to use this formula. Run the modified program for several values of n_1 and n_2.

Compare the resulting values of λ with the ones obtained with the first version of the program. Are the results identical? Are they significantly different, in your opinion?

On the following page is the flowchart for a program that starts with a number, X, takes the square root N successive times, and then resquares N successive times. If the computer carried an infinite number of significant figures, the end result would be the same number as the original X.

In reality, however, there is a small error inherent in each step of the process. The larger the number of steps (N) in the sequence, the more the errors will snowball.

Translate this flowchart into a program and experiment with various values of X and N to see at what point the final value of X begins to differ significantly from the input value of X. The results will depend somewhat on the nature of your particular computer. If you have access to two computers that differ in inherent precision or word length (say 12-bit versus 32-bit words), it would be interesting to try this program on both machines and to compare the results.

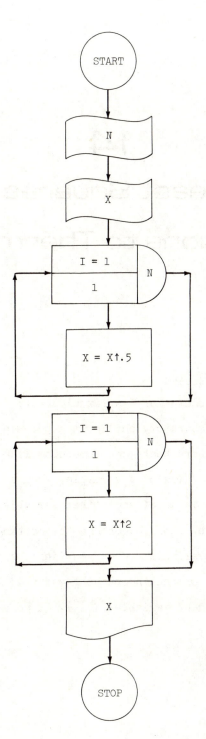

Figure 13.1 Program for taking square roots and resquaring.

14

Linear Least Squares Fitting

and Applications to Thermodynamics

14.1 Linear Least Squares Fitting

(This section employs elementary matrix algebra.)

When you go into the laboratory and do an experiment, you often find that your activities follow a pattern that goes something like this. You use some sort of apparatus to fix the value of a variable, such as temperature; then you use an instrument to measure the value of some other variable, like pressure, which depends on the first variable. Now you change the setting of the first variable and remeasure the dependent variable. When you finish such an experiment, you end up with a list of the values of the independent variable (X_i) and the corresponding observed values of the dependent variable (Y_i^{obs}). Such a list is depicted below:

$$(Y_1^{obs} \qquad X_1)$$
$$(Y_2^{obs} \qquad X_2)$$
$$(Y_3^{obs} \qquad X_3)$$
$$\cdot$$
$$\cdot$$
$$\cdot$$
$$(Y_n^{obs} \qquad X_n)$$

This list consists of "discrete" data. This means that the list contains isolated values of the variables but does not give the relationship between the variables. Thus, discrete data does not allow us to predict the Y value that would accompany some new value of X not on the list. Naturally, we would like to use our data for such prediction if possible. To do this involves finding an "analytic expression" (a formula) that expresses the relation between Y and X. That is, we want to find the function g, such that

$$Y = g(X)$$

Sometimes, it happens that the variables we are studying will have a linear relationship. This means that the relation between Y and X is of the form

$$Y = aX + b$$

where a and b are constants. To find the general relationship between X and Y, then, we need to find the value of a and b. Now, if our measurements contained no errors, we could easily find the value of a and of b by preparing a graph. We would plot our experimental data, Y versus X, on a graph and would see that the points define a straight line, as in Figure 14.1a. We would then draw a straight line passing through all the points. The value of a is equal to the slope of this line and the value of b is equal to the Y intercept. Usually, however, errors in measurement cause the experimental data to be somewhat scattered, as in Figure 14.1b.

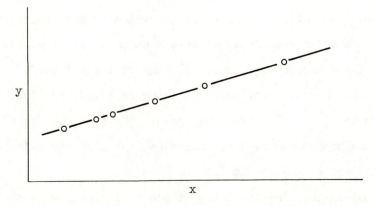

Figure 14.1a Errorless experimental data giving a linear graph.

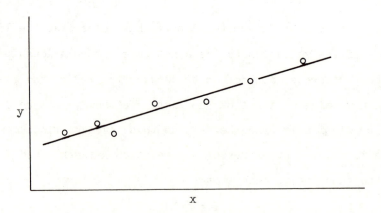

Figure 14.1b Experimental data showing scatter about a straight line.

In this case, how can you know where to draw the straight line? Obviously no straight line can pass through all the points. If you handed rulers to five different people and asked them to draw the "best" straight line through the points, you'd probably get five different lines, because each person's judgment about where to put the line would be a little different.

The technique of linear least squares fitting provides a standard method for finding the "best" straight line through a group of points. No judgment is involved in applying this method, so five different people would all arrive at the same line through a given set of points.

Let us consider our original list of data. We wish to find the value of a and of b corresponding to the "best" straight line, $Y = aX + b$, through the set of data points. Now, for each particular value, X_i, of the independent variable, we have an <u>observed</u> value, Y_1^{obs}, of the dependent variable. Using the same X_i, the straight line equation will give a <u>calculated</u> value of the dependent variable

$$Y_i^{calc} = aX_i + b$$

If the straight line actually intersects a particular data point, then $Y_i^{calc} = Y_i^{obs}$ for that point; but for real data, we have seen that the line will not usually intersect all the data points. Thus, for many of the points, $Y_i^{calc} \neq Y_i^{obs}$. Naturally, the farther the data point is above or below the line, the larger will be the difference $Y_i^{obs} - Y_i^{calc}$ for that point. To find the "best" straight line through the data points, we need to determine that line for which these differences or "deviations" are as small as possible.

To accomplish the minimization of these deviations, we define a quantity, R, called the "residual"

$$R = \sum_{i=1}^{n} (Y_i^{obs} - Y_i^{calc})^2$$

You can see that R is the sum of the squares of the deviations for all the data points. Now, we recall that $Y_i^{calc} = aX_i + b$, so that

$$R = \sum_{i=1}^{n} (Y_i^{obs} - aX_i - b)^2$$

Thus, our task is simply to find the value of a and b for which R is a minimum; hence, the name "least squares". Mathematically, we impose the minimization condition by requiring that

$$\frac{\partial R}{\partial a} = 0$$

and

$$\frac{\partial R}{\partial o} = 0$$

that is, R must be at a minimum with respect to the choice of values for a and b.

If we work out each of these partial derivatives and equate them to zero, we will obtain two linear equations in which a and b are unknowns. These equations are

$$a \sum_{i=1}^{n} X_i^2 + b \sum_{i=1}^{n} X_i = \sum_{i=1}^{n} X_i Y_i^{obs}$$

$$a \sum_{i=1}^{n} X_i + bn = \sum_{i=1}^{n} Y_i^{obs}$$

where n is the number of data points. These equations can be solved by hand using the usual method of eliminating one unknown by substitution. This approach leads to rather cumbersome formulas for a and b.

A much neater solution can be achieved by using matrices. The features of BASIC for dealing with matrices make this approach especially attractive.

In matrix form, we can write the above equations as

$$\begin{pmatrix} \sum_{i=1}^{n} X_i^2 & \sum_{i=1}^{n} X_i \\ \sum_{i=1}^{n} X_i & n \end{pmatrix} \begin{pmatrix} a \\ b \end{pmatrix} = \begin{pmatrix} \sum_{i=1}^{n} X_i Y_i^{obs} \\ \sum_{i=1}^{n} Y_i^{obs} \end{pmatrix}$$

or $\overline{M}\ \overline{P} = \overline{N}$; notice that all the elements of $\overline{M}$ and $\overline{N}$ can be calculated from the

experimental data. The solution of our pair of equations is the matrix $\overline{P}$. To find
this matrix, we left-multiply both sides of the matrix equation by the inverse of
matrix $\overline{M}$:

$$\overline{M}^{-1}\ \overline{M}\ \overline{P} = \overline{M}^{-1}\ \overline{N}$$

so that

$$\overline{P} = \overline{M}^{-1}\ \overline{N}$$

To program this procedure, we need to dimension four matrices,

$$100 \quad DIM\ M[2,2], P[2,1], N[2,1], Z[2,2]$$

(In versions of BASIC in which vectors are assumed to be column vectors, $\overline{P}$ and $\overline{N}$
can be dimensioned as vectors $P(2)$ and $N(2)$.)

The program will have to calculate, using our list of experimental data,

$$M(1,1) = \sum_{i=1}^{n} X_i^2$$

$$M(1,2) = M(2,1) = \sum_{i=1}^{n} X_i$$

$$M(2,2) = n$$

$$N(1,1) = \sum_{i=1}^{n} X_i Y_i^{obs}$$

$$N(2,1) = \sum_{i=1}^{n} Y_i^{obs}$$

This group of tasks can be accomplished compactly by the following program segment:

```
300    MAT M=ZER
310    MAT N=ZER
320    REM  N IS NUMBER OF DATA POINTS
330    READ N
340    FOR I=1 TO N
350    REM  DATA IS ORGANIZED AS X,Y PAIRS
360    READ X,Y
370    LET M[1,1]=M[1,1]+X↑2
380    LET M[1,2]=M[1,2]+X
390    LET N[1,1]=N[1,1]+X*Y
400    LET N[2,1]=N[2,1]+Y
```

```
41Ø   NEXT I
42Ø   LET M[2,2]=N
43Ø   LET M[2,1]=M[1,2]
```

Once the matrices $\overline{M}$ and $\overline{N}$ have been constructed by the above sequence, our problem can be readily finished:

```
5ØØ   MAT Z=INV(M)
51Ø   MAT P=Z*N
52Ø   LET A=P[1,1]
53Ø   LET B=P[2,1]
```

The "best" straight line through the given data points is then $Y = AX + B$, where A and B have been found in lines 520 and 530; A is the slope of this line and B is the Y intercept.

(Caution: There is a temptation when using a computer to lose touch with one's data. Never feed data blindly into a computer program without thinking about the general character of the data. The least squares procedure outlined above will fit a straight line through a set of data points whether the data is linear or not. Thus, one might be tempted to accept a "best" straight line when the data actually describes a curve. More sophisticated least squares programs usually have features that check to see how good a fit the "best" straight line gives. However, there is no substitute for an intelligent examination of the data using your eyes, your brain, and a pencil before approaching the computer.)

Problem 14.1

Complete the least squares program discussed above. Your program should take as data the number of points to be fitted and the (X, Y) values for each point. The output should be the slope (A) and Y intercept (B) of the least squares line through the points. (Try setting up a PRINT statement to output the actual equation of the line.)

Test your program on the following data:

X	Y
3.6	2.35
4.6	5.98
5.6	9.53
6.2	11.71
6.9	14.28
7.6	16.74
8.6	20.35

The least squares line through these points is Y = 3.59928X - 10.5984.

Problem 14.2

Use your program from the previous problem to fit the best straight line through the following points

X	Y
0.55	18.2
1.6	6.25
2.9	3.45
3.4	2.94
6.7	1.49
10.2	0.980
17.4	0.575
25.3	0.395

Now plot the points by hand on a piece of graph paper. On the same axes, construct the least squares line given by your program. Does the "best" line through this data give a good representation of the data?

14.2 Applications of Least Squares Fitting

We have described linear least squares fitting as a type of search for an analytic relationship between experimental variables. In this context, we were interested in the slope and intercept of the straight line only because they allowed us uniquely to identify the best line through the data points. There are situations, however, where the slope and/or the intercept of a least squares line supply us with important physical information.

(All temperatures in the following sections are assumed to be in degrees Kelvin.)

Vapor Pressure and Heat of Vaporization. We know from experience that the vapor pressure of a liquid increases as the temperature is raised. From the Clausius-

Clapeyron equation, it can be shown that the vapor pressure of a liquid is related to the temperature by this relationship:

$$\ln P = -\frac{\Delta H_{vap}}{RT} + B$$

where ΔH_{vap} is the heat of vaporization, R is the gas constant, and B is a constant of integration that depends on the pressure units chosen. A formidable equation? Not really. If we think of $(\ln P)$ as the dependent variable Y and $\frac{1}{RT}$ as the independent variable X, then the equation becomes

$$Y = -\Delta H_{vap} X + B$$

This is just the equation of a straight line with slope equal to $(-\Delta H_{vap})$. This means that if, experimentally, you determine the vapor pressure P_i of a liquid at various temperatures T_i, you can fit a least squares line through the points $(\ln P_i, \frac{1}{RT_i})$ and the value of the <u>slope</u> of the "best" line will be equal to $(-\Delta H_{vap})$.

Problem 14.3

Modify and use your program from Problem 14.1 to find ΔH_{vap} for water. Use the vapor pressure data given in the table in Problem 8.11. Note that your program cannot use the (P,T) data directly but will have to compute $\ln (P)$ for the dependent variable and $\frac{1}{RT}$ for the independent variable. The temperatures will have to be converted to degrees Kelvin. For R, use R = 1.987 cal/degree mole.

Once you have ΔH_{vap}, you can also find ΔS_{vap} since

$$\frac{\Delta H_{vap}}{T_{boiling\ point}} = \Delta S_{vap}$$

<u>Rate Constants and Activation Energy</u>. Most chemical reactions occur appreciably faster when the temperature is raised. The Arrhenius theory attempts a quantitative account for this phenomenon by specifying that the rate constant k for a reaction is

$$k = A\ e^{-\Delta E_{act}/RT}$$

The ΔE_{act} in this equation is the "activation energy." This quantity is supposed to be the amount of energy necessary to "activate" a mole of reactant to the point where the molecules will undergo reaction. The activation process is frequently thought of as leading to an intermediate "activated complex" which then

rearranges into the stable products of the reaction. ΔE_{act}, then, is the difference in energy between the activated complex and the reactant. This model has been
very useful in understanding the rate behavior of chemical reactions.

Of particular interest is the activation energy, ΔE_{act}, and its relation to
the mechanisms by which reactions proceed. Because the activated complex is a
transient entity, it might seem impossible to measure ΔE_{act}. By clever use of
least squares fitting, however, we can extract this quantity from experimental
measurements of rate constants. Observe:

$$k = A\, e^{-\Delta E_{act}/RT}$$

$$\ln k = \ln A - \Delta E_{act}/RT$$

Thus, if we measure the rate constant at various temperatures and fit a least
squares line through the points $(\ln k_i, \frac{1}{RT_i})$, the slope of that line will be equal
to $(-\Delta E_{act})$.

Problem 14.4

Suppose ΔE_{act} for a reaction is 20 kcal/mole. Write a program to determine
the rate constant for this reaction at various temperatures. Use your program to
determine the rate constant, relative to the rate constant at $0°C$, at the following
temperatures: $0°$, $10°$, $20°$, $30°$, $40°$, $50°$, $60°$, $70°$, $80°$, $90°$, $100°C$.

Problem 14.5

Use the program from the previous problem to determine the rate constant,
relative to the rate constant at $800°C$, at the following temperatures: $800°$, $810°$,
$820°$, $830°$, $840°$, $850°$, $860°$, $870°$, $880°$, $890°$, $900°C$.

Problem 14.6

There is an old rule of thumb which states that the rate of a reaction about
doubles when the temperature is raised by $10°C$. Use the results of Problems 14.4
and 14.5 to test this adage.

Equilibrium Constants and Thermodynamic Data. In assessing the character of a
chemical reaction, two thermodynamic quantities are of paramount significance.
These are $\Delta H°$ and $\Delta S°$. The first of these is the amount of heat given off or

absorbed by the reaction when carried out under standard conditions. If ΔH^O is
negative, heat is given off; if positive, heat is absorbed. The second quantity,
ΔS^O, is the change in molecular order. If ΔS^O is positive, the products of the
reaction represent a more disordered assemblage than the reactants; if ΔS^O is negative, the reaction results in an increase in molecular order.

Both ΔH^O and ΔS^O for a reaction can be determined by measuring the equilibrium
constant for the reaction at various temperatures.   It can be shown that

$$\Delta G^O = -RT \ln K$$

where K is the equilibrium constant and ΔG^O is the standard free energy change for
the reaction. We know from basic thermodynamics that

$$\Delta G^O = \Delta H^O - T \Delta S^O$$

at any particular temperature. Then

$$\Delta H^O - T \Delta S^O = \Delta G^O = -RT \ln K$$

and

$$\ln K = - \frac{\Delta H^O}{RT} + \frac{\Delta S^O}{R}$$

Notice that, if $\ln K$ is considered to be the dependent variable Y and $\frac{1}{RT}$ is the
independent variable X, then we have

$$Y = -\Delta H^O X + \frac{\Delta S^O}{R}$$

This is just the equation of a straight line with slope equal to $(-\Delta H^O)$ and Y
intercept equal to $\frac{\Delta S^O}{R}$. Again, the straight line displays its analytic power. If
we measure K_i for a reaction at several temperatures T_i and fit a least squares
line through the points $(\ln K_i, \frac{1}{RT_i})$, then $-\Delta H^O$ for the reaction is equal to the
slope of this line and ΔS^O is equal to the Y intercept multiplied by R.

Problem 14.7

The equilibrium constant for the gas phase reaction

$$N_2 + 3H_2 \rightleftharpoons 2NH_3$$

is quite temperature dependent. Some values of the equilibrium constant (in terms
of concentration) are as follows:

	Temperature	
T, °C		K
350		1.79
400		0.492
450		0.151
500		0.0575

Use your least squares fitting program to determine ΔH^{o} and ΔS^{o} for this reaction. Can you explain the sign of ΔH^{o} and of ΔS^{o} in physical terms?

Answers to Selected Problems

There is room for individual style in writing your computer programs. However, the results given by your programs must be correct. The answers provided here will allow you to check some of your programs to be sure they are working properly. It is always advisable to test a newly written program before using the program for "production" work.

CHAPTER 6 Problems

6. The two scales coincide at -40°

7. The (111) planes have a spacing of 2.059 Å

8.

N	$\log_{10}(N)$	$\log_e(N)$
1	0	0
2	0.3010	0.6931
3	0.4771	1.0986

etc.

CHAPTER 7 Problems

1a. 12 g nitrogen contain 5.161×10^{23} atoms

2a. 3.25 moles of helium has a mass of 13.01 g

3a. 46.45 percent lithium; 53.55 percent oxygen

4a. 92.83 percent lead; 7.17 percent oxygen

5a.   69.55 percent nitrogen; 30.45 percent oxygen

6a.   46.45 percent lithium; 53.55 percent oxygen

8a.   Compound is FeS

9a.   Compound is FeO

10a. Compound is C_6H_6

11a. Compound is $MgCl_2$

12a. Compound is $HgCl_2$

CHAPTER 8 Problems

1b.   Volume at STP is 0.3421 liters

7.   At $700^\circ K$, for CO_2

P(ideal)	P(van der Waals)	Volume
718 atm	982 atm	0.08 liters
638 atm	773 atm	0.09 liters

etc.

11b. Volume of dry gas at STP is 0.08934 liters

17.   Velocity of HCl relative to velocity of H_2 is 4.253/1

CHAPTER 9 Problems

1.   Resulting $CaCl_2$ solution is 0.1391 molar

4.   The 2.136 molar alcohol solution is 2.412 molal

6.   Molecular weight of substance 1 is 180 g/mole

7.   With 1 qt of antifreeze, the engine will freeze at $22.0^\circ F$ ($-5.54^\circ C$)

CHAPTER 10 Problems

2a. $[SO_3]$ = 0.9787 mole/liter; $[SO_2]$ = 0.02131 mole/liter; $[O_2]$ = 0.5107 mole/liter

3a. $[NH_3]$ = 0.2091 mole/liter; $[N_2]$ = 0.8955 mole/liter; $[H_2]$ = 0.6864 mole/liter

7.   Solubility of CaF_2 is 2.154×10^{-4} mole/liter

8.   pH of 0.05M oxalic acid solution is 1.48

10e. pH of 0.1M acetic acid solution is 2.87

11a. pH of 0.001M phosphoric acid solution 3.06

15a. pH of 0.001M Na_3PO_4 solution is 11.0

CHAPTER 11 Problems

1. Decay constant for ^{90}Sr is 0.02476 years^{-1}

2. For ^{90}Sr

Amount left	Time elapsed
10.0 g	0 yr
9.76 g	1 yr
8.84 g	5 yr
7.81 g	10 yr

etc.

3. The Dead Sea scroll fragment is about 1,960 years old

CHAPTER 12 Problems

6. The distance between the carbon atom and atom N1 is 1.33$\overset{o}{\text{A}}$

CHAPTER 14 Problems

3. The enthalpy of vaporization of water at 100°C is 9.717 kcal/mole
    (literature value)

4.

Rate relative to rate at 273.15^{o}K	T, ^{o}K
1	273.15
3.67	283.15
12.4	293.15

Appendix 1

This appendix presents an abbreviated list of system commands for the Hewlett-Packard 2ØØØ series time-sharing systems. We recommend that the user of other systems check with his computer center to get a list and interpretation of the comparable system commands.

The individual system commands are shown in both their full and abbreviated forms; i.e., their first three letters are sufficient. When specific statement numbers are used in an optional manner, the system will begin with the statement specified (if it exists) or the next higher (if it does not exist), and will terminate with that statement (if it exists) or the next lower (if it does not exist). This relates specifically to DELETE, LIST, PUNCH, and RUN.

Full Name	Example of Use	Purpose
APPEND	APPEND-PROG APP-PROG	Appends (adds) the named program, PROG, to the program in the user space.
BYE	BYE	Logs user off the system; i.e., ends the session on the terminal.
CATALOG	CATALOG CAT	Lists the names and corresponding lengths of all programs in the user's library.
DELETE	DELETE-1ØØ DEL-1ØØ	Deletes (erases) all program statements after and including the one specified.

Full Name	Example of Use	Purpose
DELETE	DELETE-100,200 DEL-100,200	The same as above for those statements from the first statement number to the second, inclusively.
ECHO	ECHO-OFF ECH-OFF	After sign on, use of the command will permit the user to operate terminal in half duplex (one wire) mode of operation.
	ECHO-ON ECH-ON	Returns the computer to full duplex (two wire) mode of operation.
GET	GET-NED GET-$PUP	Copies program from user library to user space when used without $ preceding program name, and from system library to the user space when used with the $.
HELLO	HELLO-Z115,SSC HEL-Z115,SSC	Log on allows new terminal session to begin. User must have a valid I.D. code (Z115) and password (SSC).
KILL	KILL-NED KIL-NED	Allows user to delete a program from the user's library. NOTE: A program must be KILLed before a new version may be stored.
LENGTH	LENGTH LEN	Allows the user to find the number of words in his current program.
LIBRARY	LIBRARY LIB	Lists the names and lengths of all programs in the system library.
LIST	LIST LIS	Allows the user to have the terminal print out the program in the user space.
	LIST-100 LIS-100	The same as for LIST, but beginning with the line specified.
	LIST-100,200	The same as for LIST-100, but printing up to and including line 200.
NAME	NAME-PUP NAM-PUP	Assigns the characters after the - as the name of the program in the user space. It is used to identify a program for future retrieval with a GET command, and may be any combination of from one to six printing characters. However, the first character may not be either a $ or an *, and the name may not contain any commas or quotation marks.
PUNCH	PUNCH PUN	Allows the user to have the terminal punch a copy on paper tape of the program in the user space to be saved and reentered later. NOTE: This command should be used and not LIST, as LIST tapes will have missing control characters and may not be reloadable.

Full Name	Example of Use	Purpose
PUNCH	PUNCH-1ØØ PUN-1ØØ	The same as above, but beginning with a specified line.
	PUNCH-1ØØ,2ØØ PUN-1ØØ,2ØØ	The same as above, but beginning and ending with a specified line number.
RENUMBER	RENUMBER REN	Renumbers the program in the user space assigning the line number 1Ø to the first line of the program and all subsequent line numbers in increments of 1Ø. NOTE: All BASIC commands specifying a branch or change of control will be modified properly also.
	RENUMBER-4Ø REN-4Ø	The same as above, but beginning with the line specified.
	RENUMBER-4Ø,I REN-4Ø,I	The same as above, but the increment to be used is I, not 1Ø.
RUN	RUN	Causes the program to begin execution.
	RUN-5Ø	The same as above, but beginning at line 5Ø.
SAVE	SAVE SAV	Causes a copy of the program in the user space to be stored in the user's library.
SCRATCH	SCRATCH SCR	Erases the program in the user's work space. NOTE: It does not erase the name from the user space.
TAPE	TAPE TAP	Informs the system that the following input is on punched paper tape.
TIME	TIME TIM	Causes the terminal to print out the current port number and time to date since last reset of password, usually the first of the month.

Appendix 2

This appendix presents a comparative chart of six versions of the BASIC language. The six language versions presented here are: the original Dartmouth BASIC, the Hewlett-Packard 2000A BASIC, Digital Equipment Corporations PDP-11 BASIC, the Honeywell 1640 BASIC, IBM's CALL/360 BASIC, and the General Electric 635 BASIC. The language elements are identified in the left-most column of the chart, and the six columns on the right-hand side of the chart present element usage in each of the six BASIC languages, identified respectively by the column headings Dartmouth, H-P 2000A, PDP-11, H1640, Call/360, and GE 635.

Language Element	Dart- mouth	H-P 2000A	PDP- 11	H 1640	Call/ 360	GE 635
Arithmetic Operators:						
Addition	+	+	+	+	+	+
Subtraction	-	-	-	-	-	-
Multiplication	*	*	*	*	*	*
Division	/	/	/	/	/	/
Exponentiation	↑	↑	↑	↑ or **	**	↑
Relational Operators:						
Less than	<	<	<	<	<	<
Less than or equal to	<=	<=	<=	<= =<	<=	<=
Greater than	>	>	>	>	>	>
Greater than or equal to	>=	>=	>=	>= =>	>=	>=
Equal to	=	=	=	=	=	=
Not equal to	<>	<> #	<>	<> ><	<>	<>

243

Language Element	Dart-mouth	H-P 2000A	PDP-11	H 1640	Call 360	GE 635
Basic Functions:						
SIN(X)	X	X	X	X	X	X
COS(X)	X	X	X	X	X	X
TAN(X)	X	X	X	X	X	X
ATN(X)	X	X	X	X	X	X
EXP(X)	X	X	X	X	X	X
ABS(X)	X	X	X	X	X	X
LOG(X)	X	X	X	X	X	X
SQR(X)	X	X	X	X	X	X
SGN(X)	X	X	X	X	X	X
RND(X)	X	X	X	X	X	X
INT(X)	X	X	X	X	X	X
COT(X)				X		
Matrix Commands:						
MAT A=B	X	X	X		X	X
MAT C=A+B	X	X	X		X	X
MAT C=A-B	X	X	X		X	X
MAT C=A*B	X	X	X		X	X
MAT C=(k)*B	X	X	X		X	X
MAT C=INV(A)	X	X	X		X	X
MAT C=TRN(A)	X	X	X		X	X
MAT C=ZER	X	X	X		X	X
MAT C=CON	X	X	X		X	X
MAT C=IDN	X	X	X		X	X
MAT READ A	X	X	X		X	X
MAT PRINT B	X	X	X		X	X
MAT INPUT			X		X	
Logic Commands:						
FOR I=A TO B STEP C NEXT I	X	X	X	X	X	X
GOSUB RETURN	X	X	X	X	X	X
GO TO	X	X	X	X	X	X
GO TO I OF ...		X				
ON I GO TO ...				X	X	X
RESTORE	X	X	X	X	X	X
IF ... THEN	X	X	X	X	X	X
IF ... GO TO				X	X	X
STOP	X	X	X	X	X	X
END	X	X	X	X	X	X
REM	X	X	X	X	X	X
Data Commands:						
DIM A(a),B(a,b)	X	X	X	X	X	X
DATA a,b,c ...	X	X	X	X	X	X

Language Element	Dart- mouth	H-P 2ØØØA	PDP- 11	H 164Ø	Call 36Ø	GE 635
Assignment Commands:						
LET	X	X	X	X	X	X
Optional LET		X		X	X	
DEF FN	X	X	X	X	X	X
Input/Output Commands:						
INPUT	X	X	X	X	X	X
READ	X	X	X	X	X	X
PRINT	X	X	X	X	X	X

Index

246